Randomness, Statistics, and Emergence

SECOND EDITION

Philip McShane

M.Sc., Lic. Phil., S.T.L., D. Phil. Oxon.

edited by
James Duffy and Terrance Quinn

AP

AXIAL PUBLISHING
Vancouver

Axial Publishing
2-675 Victoria Drive
Vancouver, British Columbia
V5L 4E3 Canada
www.axialpublishing.com

Canadian Cataloguing in Publication Data
McShane, Philip, 1932–2020

ISBN   978-1-988457-08-6
1. Methodology 2. Philosophy of Science I. Title

Text layout and cover:
Mayra Andrea Tejeda Chávez
James Duffy

To my mother
and the memory
of my father

Table of Contents

A rare plant has reappeared after more than a century in hiding. The pinkish-flowered plant, known as grass-poly, was found growing on the banks of an old farmland pond in Norfolk [England].

The mystery species "came back from the dead" after seeds submerged in the mud were disturbed during work to restore the pond.

. . .

"It's really quite beautiful," says Prof Sayer. "We only found a handful of these plants in the pond but we're hoping to cultivate this population and keep it going and expand it now we know it's there."[1]

Preamble

Philip McShane's book *Randomness, Statistics and Emergence* was first published in 1970.[2] Copies are available in the Bodleian and other university libraries. For about half a century, however, it has mainly lain, as it were, undisturbed. That is regrettable since *RSE* is a remarkable book that has the potential for contributing to progress in philosophy of science and science, globally. And so, it is with great pleasure that I take this opportunity to provide an introduction to this second edition of the book.

You may wonder why *RSE* has not yet featured in the mainstream literature. There are many examples in history where important works at first were missed by the scientific community. In the case of *RSE*, a main difficulty is that the method employed to write the book, and the method required to read it, are discontinuous with the ethos of modern philosophy of science. In a tradition that goes back to and includes the works of Bernard de Fontenelle (1657–1757), modern philosophy of science mainly debates models defined in general terms remote to scientific practice and concrete situations.[3] The approach advocated by McShane is quite different. It asks that one advert to one's experience in scientific practice and that way obtain detailed accounts of science, concretely, in verifiable terms.

---

[1] Helen Briggs, "Surprise discovery of rare plant at Norfolk 'ghost pond,'" *BBC News, Science*, November 28, 2020, https://www.bbc.com/news/science-environment-55104153, (accessed December 1, 2020).

[2] Philip McShane, *Randomness, Statistics and Emergence*, Dublin and London, Gill and Macmillan and Macmillan, 1970. From here on, I refer to the book as *RSE*.

[3] Regarding Fontenelle's work, see H. Butterfield, *The Origins of Modern Science, 1300–1800*, New York, G. Bell and Sons Ltd., 1957, pp. 164–167.

[*RSE*] is an attempt to orientate philosophy of science away from general considerations towards a reflection on the detailed content of science and on the details of procedure in scientific investigation. So, for example, in chapter eight, instead of dealing in general terms or in terms of philosophic positions with the problems of the foundations of statistics, we deal with the content and expression of some standard work in probability theory. For, the philosopher of science has the task of transforming and correctly orientating science from within, and he can do so adequately only in so far as he includes as data for philosophic reflection the details of contemporary science.[4]

And so, "the approach is of a piece with the philosophic method, which is one of attention to oneself in the performance of knowing, because it can only be established thematically through the subject's attention to his or her own performance in knowing."[5]

But what are "reflection on the detailed content of science and on the details of procedure in scientific investigation"[6] or "attention to oneself in the performance of knowing"?[7] These will sound odd or implausible to most contemporary philosophers and will be of dubitable interest to scientific investigators. And yet, what is your view or model of scientific knowing? Does it bear out with your experience, in instances, in detail, in, for example, physics, biology, or ecology?

The method described by McShane is not a model in the sense of providing an answer or prescribing rules. It is, however, an approach that regards questions in scientific practice, including questions about questioning. And so, for instance, Feyerabend's followers cannot entirely dismiss the possibility of such a method without performance-contradiction, for their view regards scientific inquiry. But that is just an aside. The method advocated by McShane is something that needs to be tried. In that way, for instance, progress in one's view of progress in physics will be based on what one finds by adverting to and drawing on one's own experience in progress in physics. That can sound innocuous enough. But asking scholars to advert to their experience in scientific practice is not yet part of either the philosophic or scientific traditions. And to do so collaboratively and with precision needed to handle problems of our times will not be easy.[8]

Bernard Lonergan provides a brief preliminary sketch of a few aspects of the methodological development anticipated:

---

[4] *RSE* (1970), vii; page lxiii below.

[5] *RSE* (1970), 13; page 10 below.

[6] See note 4.

[7] See note 5.

[8] Eventually, effective collaboration will be "functional." See note 38 and section 4.3.

As science develops, philosophy is impelled to migrate from the world of theory and to find its basis in the world interiority. ... It is not enough to have acquired common sense and to speak ordinary language. One has to examine mathematics and discover what is happening when one is learning it and, again, what was happening as it was being developed. From reflecting on mathematics one has to go on to reflecting on natural science, discern its procedures, the relations between successive steps, the diversity and relatedness of classical and statistical methods, the sort of world such methods would reveal—all the while attending not merely to scientific objects but also attending, as well as one can, to the conscious operations by which one intends the objects. From the precision of mathematical understanding and thought and from the ongoing, cumulative advance of natural science, one has to turn to the procedures of common sense, grasp how it differs from mathematics and natural science, discern its proper procedures, the range of its relevance, the permanent risk it runs of merging with common nonsense. To say it all with the greatest brevity: one has not only to read *Insight* but also to discover oneself in oneself.[9]

Lonergan's book *Insight*, however, is extraordinarily dense and advanced in its reach. Fortunately, help is available in *RSE*. McShane's book invites readers to engage in and make beginnings in sorting out details in a strategically ordered, cumulative[10] series of key problems in modern science and philosophy of science. This is not to suggest that *RSE* is a departure from Lonergan's heuristics. McShane's book is the kind of work called for by *Insight*: "[O]ne reaches familiarity with the notion of insight by modifying the illustrations."[11] And so, according to McShane, "[t]he world view that should emerge—in the reading subject—is the world view of *Critical Existentialism*, a view which originated with Bernard Lonergan."[12]

This introduction to *RSE* has four sections.

---

[9] *Method in Theology*, New York, Herder and Herder, 1972, 260; *Method in Theology*, Robert M. Doran and John D. Dadosky (eds.), vol. 14, Collected Works of Bernard Lonergan, Toronto, University of Toronto Press, 2017, p. 244.

[10] *RSE* (1970), 13; page 10 below.

[11] *Insight: A Study of Human Understanding*, vol. 3, Collected Works of Bernard Lonergan, Frederick Crowe and Robert Doran (eds.), Toronto, University of Toronto Press, 1992, p. 56.

[12] *RSE* (1970), ix; page lxiv below. This is part of a longer quotation given in section 3. See note 73 on pages xiii–xiv below.

In section 1, I recall my first readings of *RSE*, 1993–1995. The book helped me make progress in foundations[13] and helped me begin to see philosophy of physics with a more discerning eye.

In section 2, I touch on historical contexts, 1970–2020. I point to a few works that cite or make use of *RSE*. At the same time, not attempting historical analysis but merely by way of illustration, I also include an example of philosophical reflection on scientific progress given by a leading contemporary scholar.[14]

In the 27 years since I first encountered *RSE*, and with the help of McShane's subsequent works,[15] I have gradually been making progress on various fronts, including in the metaphysics of chapters 15ff of *Insight*. I do not mean to suggest anything like "mastery." My growth—such as it has been—mainly has been deepening my realization that *Insight* maps developmental sequences that are remote to present-day achievement.

In section 3, I call attention to what in my present (2020) view are some key advances made in *RSE* in 1970. McShane's results are based on his adverting to his experience in scientific practice. By lending support to the book as I do, I reveal something of my own experience, views, and method. As alluded to in the text at note 5, determining whether or not the book's method and results should be expanded on (because they might have the potential to contribute to progress in philosophy and science, globally) will not be settled by traditional philosophic debate. In order to evaluate the book (method and results), one needs to make progress in implementing the method employed in the book, in instances, and in detail. On the other hand, if one chooses to "skip the exercises," one will be missing out on the possibility of invited growth; and "arguments ... will appear inconclusive."[16]

---

[13] "Foundations may be conceived in two quite different manners. The simple manner is to conceive foundations as a set of premises, of logically first propositions. The complex manner is to conceive of foundations as what is first in any ordered set. If the ordered set consists in propositions, then the first will be the logically first propositions. If the ordered set consists in an ongoing, developing reality, then the first is the immanent and operative set of norms that guides each forward step in the process." CWL 14, 253. In the present context, I am referring to the "ongoing, developing reality."

[14] Alexander Bird, "Scientific Progress," in Paul Humphreys (ed.), *The Oxford Handbook of Philosophy of Science*, Oxford, Oxford University Press, 2016, pp. 544–563. https://www.oxfordhandbooks.com/view/10.1093/oxfordhb/9780199368815.001.0001/oxfordhb-9780199368815-e-29, (accessed December 1, 2020).

[15] See note 21.

[16] *RSE* (1970), 13; page 10 below.

Now,

[a]ny worthwhile philosophic position is open to the possibility of …
contraction, and a standard manifestation of such contraction is the
summary. The summary can give the impression of capturing the
essence of a position. But a summary expresses the essence only in so
far as the summarizer has the essence of the position in [their] mind.
In this respect one may note that the book *Insight* is a summary
expression of a philosophic position. As such it provides a phantasm
for the reader which requires elaborate supplementation if the reader
is to reach the mind of the author.[17]

These words also apply to *RSE*. And so, while section 3 touches on a range
of results from the book, it is not intended as "summary." It is merely a series of
pointings. The section is intended to intimate something of the scope and control
of meaning[18] that can be obtained through "self-attention" in scientific practice.
If, on a first reading of section 3, you find some of the claims from *RSE*
surprising, I hope that you reserve judgment. As alluded to above, meanings and
grounds of results will emerge gradually, by working through the book. Another
purpose of section 3, then, is to raise interest in reading *RSE*; and to encourage
readers to do so in the manner intended by the author.[19]

There is, however, a still further purpose of section 3, namely, to suggest
that if one enters into the detailed and cumulative[20] exercises in self-attention,
one will find sufficient evidence to reach a decision: Yes, for the sake of progress,
we should take up *RSE*'s method and move forward with its results.

What might that "moving forward with its results" look like? One way to
answer that is to look to subsequent writings of Philip McShane (1932–2020).[21]
The opera omnia of McShane reveals a remarkable and accelerated growth
trajectory. The "genetic"[22] series in methods and results reached by him, de facto,
resulted from *RSE*.[23] My present focus, however, is different. It is immediate, and
personal. I have learned much from *RSE,* and I continue to find it a valuable
resource. Keeping paws, poise and focus on the book, are there *contemporarily*

---

[17] *RSE* (1970), viii; page lxiv below.

[18] CWL 3, 530; and entries in Index under 'meaning,' control of. This can also be
called being "luminous" in one's performance in science.

[19] *RSE* (1970), 13; page 10 below. See notes 4 and 5.

[20] *RSE* (1970), 13; page 10 below.

[21] Most of McShane's academic writings are available on his website,
http://www.philipmcshane.org and through Axial Publishing,
http://www.axialpublishing.com, (accessed December 1, 2020).

[22] CWL 3, 609–10.

[23] Much of McShane's work of the next fifty years was his following up on issues
raised in *RSE*. See note 21. See also sections 3.4 and 4, below.

significant results or directions that, within *RSE* itself, are burgeoning or to some extent imminent? Can I nudge those into view? What are some aspects of "the view that would result from developing"[24] the method and results of *RSE*? Section 3.4 provides a short-list of results in *RSE* that, in my present (2020) reading of the book, call out for some such development. For reasons given at the beginning of section 4, I select three of these for discussion.

RSE brings out that "schemes of recurrence (or recurrence-schemes)"[25] are "units of evolution"[26] of world process, identified to be "emergent probability."[27] A technical result in heuristics is that, once a scheme of recurrence is "concretely possible," the (empirical) probability of emergence[28] will be something like a sum of probabilities of constituent events. Investigators familiar with methods for statistical analysis of "cycles" in biology or ecosystems may well find this heuristics obvious or at least plausible. But for many readers, more detail would be helpful. Without attempting to go beyond "pre-formulational" aspects of the problem, [29] section 4.1 is on heuristics of probabilities of emergence of recurrence-schemes.

Where section 4.1 regards (emergence of) individual schemes, section 4.2 looks to the problem of handling many schemes at once. As mentioned above, I write from foundations[30] that have emerged from my growth in heuristics of "Space and Time"[31] which, in turn, have been part of my reach into chapters 15ff of *Insight* (and more) and McShane's works. Sometimes investigators focus on events, such as in "particle event-oriented" research at CERN. [32] In other contexts, inquiry is "multi-event-oriented" or even "multi-cycle-oriented," such as in ecology and evolutionary studies. In order to develop practical heuristics for investigating "conjoined pluralities" of recurrence-schemes, new control of meaning will be needed in identifying constituent "aggreformic" events. But also needed will be new heuristics and mathematical structures for handling "complexly overlapping sets" of recurrence-schemes.

---

[24] *Method in Theology*, 250; CWL 14, 235. Indicating the view that results from one's position is a pivotal for reaching an accord on progress. See also note 262 below.

[25] *RSE* (1970), 206–229; pages 171–189 below; and CWL 3, entries in Index under 'recurrence, schemes of.'

[26] *RSE* (1970), 9–10; page 7 below.

[27] *RSE* (1970), 5–10; pages 4–7 below.

[28] *RSE* (1970), 230; pages 191 below.

[29] *RSE* (1970), 151, 155, 157; pages 125, 128, 130 below.

[30] See note 13.

[31] See, e.g., James Duffy, Robert Henman, and Terrance Quinn, "The Heuristic Notions of Space and Time," *Journal of Macrodynamic Analysis*, vol. 14, 2020, pp. 65–94.

[32] *Conseil Européen pour la Recherche Nucléaire*, https://home.cern, (accessed December 1, 2020).

Section 4.3 is for readers who are concerned by the deepening global crisis of modern times and also have made some progress toward sensing the (distant) possibility of a major shift in collaboration, anticipated by Lonergan's 1965 discovery of "functional specialization," "Lonergan's Dream."[33] In various ways, that shift is "(pre-) emergent" in history.[34] Aspects of "functional collaboration" were symbolized by McShane: $C_{ij}, i, j = 1, 2, \ldots, 9$.[35] But echoing note 3 from the second Preface, the "topic cannot be developed here." Section 4.3 then also is intended in the same spirit as section 3.2, that is, not as summary but to raise questions.

As you might infer, heuristics of sections 4.1 and 4.2 apply to human affairs in history. And so, questions arise regarding probabilities of emergence of *schemes of recurrence in functional collaboration*. Section 4.3 comments briefly (and somewhat superficially but I think helpfully) on timelines and some statistical structures that will be involved.

Present divisions in the Academy are "to be deeply lamented."[36] Science continues to make notable advances but lacks an adequate heuristics of human progress. Philosophy of science grows in sophistication and technical achievement but is dominated not by reflection on experience in scientific practice but by debates in general terms and about merely speculative models.[37] There is a tradition of scholarship inspired by Lonergan's writings but that has not *yet* been promoting foundations in scientific understanding upon which, in modern contexts, "conception, affirmation, and implementation"[38] will much

---

[33] Pierrot Lambert and Philip McShane, *Bernard Lonergan: His Life and Leading Ideas*, Vancouver, Axial Publishing, 2010, p. 163.

[34] I use the prefix to refer to the fact that while distinct tasks are emerging in history, work is not yet luminous. Lonergan's proposal is a "model" but "not merely" a model (CWL 14, 4). It is a global field theory discoverable in the "data of any sphere of scholarly human studies" (CWL 14, 336). Among many possible sources, see, e.g., P. McShane, *Futurology Express*, Vancouver, Axial Publishing, 2013, ch. 1, "The Turn-Around," pp. 7–13; and Terrance Quinn, *The (Pre-) Dawning of Functional Specialization in Physics*, Hackensack, New Jersey, World Scientific, 2017.

[35] Philip McShane, *Interpretation from A to Z*, Vancouver, Axial Publishing, 2020, p. 26; Philip McShane, *A Brief History of Tongue: From Big Bang to Coloured Wholes*, Nova Scotia, Axial Press, 1998, p. 108.

[36] CWL 14, 339. Lonergan was referring to divisions in the Christian tradition. However, as the paragraph brings out, the sentiment applies.

[37] The literature on scientific progress is vast but includes the works of, for instance, Popper, Kuhn, Salmon, Lakatos, Putnam, Jeffrey, Feyerabend, Hacking, Howson and Urbach, and many others. Bird provides a point of entry into the literature. See note 14.

[38] CWL 3, 416. Link this to the need for "resolute and effective intervention in the historical process." *Phenomenology and Logic*, vol. 18, Collected Works of Bernard Lonergan, Philip McShane (ed.), Toronto, University of Toronto Press, 2001, pp. 305–308.

depend.[39] It is my expectation, however, that readers who work through *RSE* in the manner intended by the author will find it helpful toward resolving present divisions and, as hoped by the author, toward attaining an adequate modern *Weltanschauung*.[40]

1. My first readings of *RSE*, 1993–1995

In 1992–1993, I was at the School of Mathematics, Trinity College Dublin. It was my first of three post-doctoral years. A civilized teaching schedule allowed me time to concentrate on and make new progress with the early chapters of *Insight* (with examples from science both ancient and modern). I also enjoyed an introduction to 20th century quantum field theory, in a seminar given by Lochlainn O'Raifeartaigh, at the Dublin Institute for Advanced Studies (DIAS). The seminar was on Lie algebras and symmetry methods in gauge theory.[41] It was both rich and challenging. I was not lacking a decent background in modern mathematical physics. But my graduate degrees were in operator theory, that is, the mathematics more than the physics that emerged from 20th century quantum theories.

In the fall of 1993, I started into what turned out to be a two-year position as College Lecturer at University College Cork (UCC). A long-term goal of mine was to make progress toward an up-to-date worldview that would include, among other things, science, art, philosophy, and theology. It seemed like a good idea to try my hand at a contemporary problem in philosophy of physics. I decided to look at Bell's inequality[42] and related paradoxes. Serendipitously, it was at this

---

[39] Lonergan articulated this need throughout his works. He was once asked how much science a theologian needs to know: "Well, they have to be able to read with profit a book like Lindsay and Margenau's *Foundations of Physics*," https://bernardlonergan.com/Archive, File 89000DTE070, p. 3, (accessed December 1, 2020). This is a transcript of question-and-answer session, Boston College Lonergan Workshop, June 18, 1976. The answer was not offhand. Questions were given to Lonergan in the morning. See Philip McShane, "Collective Futurology," *Divyadaan: Journal of Philosophy and Education*, vol. 28, no. 2, 2017, p. 190, n. 3. The copy that I have of the book to which Lonergan referred is Robert Bruce Lindsay and Henry Margenau, *Foundations of Physics*, New York, Dover Publications, 1957. The answer would now need to upgraded to consider current contexts. For physics, a modern counterpart to Lindsay and Margenau would be a book like Richard Healey, *Gauging What's Real: The Conceptual Foundations of Contemporary Gauge Theories*, Oxford University Press, 2007. A key to Lonergan's pointing is the phrase "to read with profit." See note 50.

[40] *RSE* (1970), ix; page lxiv below.

[41] A reference text for the seminar was Lochlainn O'Raifeartaigh, *Group Structure of Gauge Theories*, Cambridge University Press, 1988.

[42] J.S. Bell, "On the Einstein Podolsky Rosen Paradox," *Physics*, vol. 1, no. 3, 1964, 195–200; and J. Bell, "Bertlmann's Socks and the Nature of Reality," *Journal de Physique Colloques*, 1981, vol. 42 (C2), pp. 41–61.

time that I also started reading McShane's book, *RSE* (a copy of which was available to me in the UCC library).

Working through *RSE* gave me a major foundational boost.[43] The method employed in the book was not new to me. It was the same method invited by Lonergan in *Insight*. But where *Insight* is densely doctrinal (i.e., *descriptive* of work to be done and of Lonergan's achievement), *RSE* includes extensive detail. It helped me climb into historically contextualized series of exercises in self-attention, in and mediated by modern science and modern philosophy of science.

As a result, I was soon able to identify various fundamental errors running through the otherwise fascinating literature on Bell's inequality. I was also beginning to see, in instances, that much of what was written on the inequality was not about (experience in) quantum physics but was developed, instead, on the basis of speculative models at considerable remove from experience and scientific practice in quantum physics. While some arguments were intricate weaves indeed—of logic and mathematics—it became evident to me that a fundamental reorientation in method was needed.

On the positive side, working through *RSE* helped me reach a preliminary control of meaning in statistical method in various modern scientific contexts. I began to see that statistical method would be foundational to an adequate modern heuristics of world process; that "schemes of recurrence"[44] are essential to "evolutionary studies" and that identifying their significance in world process[45] was one of Lonergan's (less discussed but) important discoveries. *RSE* also provided glimpses of the fact that, eventually, an essential piece of the puzzle would be our growth and development in heuristics of growth and development.

2. Some historical contexts, 1970–2020

In the Preface to the second edition of *RSE*,[46] McShane recalls that external reviewers of his PhD thesis (the source text for the book) asked that chapter 8 be removed, on the basis that it was "merely pure mathematics."[47] I remember talking with McShane about this, in Halifax, circa 1986. It came up in conversation because the problem pertains to essential aspects of the method of the book, namely, "generalized empirical method." The method does not attempt to simplify or summarize[48] scientific results but rather seeks to be luminous in scientific results, whatever those results happen to be, mathematical or otherwise. In 1985, Lonergan provided a precise definition: "Generalized empirical method operates on a combination of both the data of sense and the data of

---

[43] See note 13.
[44] See note 25.
[45] See, e.g., CWL 3, 141–152.
[46] See pages liii–lxi below.
[47] See page lv below, note 6.
[48] See note 17.

consciousness: it does not treat of objects without taking into account the corresponding operations of the subject; it does not treat of the subject's operations without taking into account the corresponding objects." [49] As McShane wrote in 1995, "[t]he real creative dialogue is with human inquiry in developed fields. But that dialogue's fruitfulness pivots on inner dialogue of generalized empirical method. More prosaically, one cannot do philosophy of *x* without competence in *x*."[50]

Preparing to write this introduction, I remembered my 1986 conversation with McShane and then wondered how other reviewers[51] received *RSE*, near the time of its first publication. What I found was that those who were interested in Lonergan's work were mainly positive about the book while some reviewers who were scholars in philosophy of science were rather negative. Whether positive or negative, however, all the reviews concentrated on nominal content; and none appealed to scientific results or scientific practice. I mention this here not to stir up old debates but to draw attention to the fact that the method advocated (namely, "generalized empirical method"[52]) was (and still is) not standard practice.

What has happened in the half century since that initial publication of *RSE*? As alluded to in the first paragraph of section 1, in the "[fifty] years after the first publication of [*Randomness, Statistics, and Emergence*], it would seem that the development of thought on probability and statistical knowledge among mathematicians, scientists, statisticians and philosophers of science has, for the most part, bypassed any consideration of the clues presented by [McShane]."[53] In the quotation, Melchin was speaking about the 25-year period after the first publication of *Insight*. The observation, however, also applies to the 50-year period after the first publication of *Randomness, Statistics and Emergence*.[54]

---

[49] Bernard Lonergan, *A Third Collection*, vol. 16, Collected Works of Bernard Lonergan, Robert Doran and John Dadosky (eds.), Toronto, University of Toronto Press, 2017, p. 136.

[50] Philip McShane, *Economics for Everyone, Das Jus Kapital*, Edmonton, Alberta, Commonwealth Publications, 1995, 158–59. See also notes 75–79 below.

[51] Garrett Barden, *Philosophical Studies*, vol. 20, 1971, pp. 344–346; R. J. Blackwell, *The Modern Schoolman*, vol. 49, 1971-1972, p. 89; John Heywood Thomas, "Review of *Randomness, Statistics and Emergence*, by Philip McShane," *The Clergy Review*, vol. 56, 1971, pp. 310–312; W. Mathews, *The Heythrop Journal*, vol. 13, 1972, pp. 319–321; W. Newton-Smith, *Bibliography of Philosophy*, vol. 18, 1971, p. 34; P. Quay, *Review for Religious*, vol. 30, 1971, p. 30; and Denis Ryan, "Review of *Randomness, Statistics and Emergence* by Philip McShane," *The Furrow*, vol. 22, 1971, pp. 596–598.

[52] See note 49.

[53] Ken Melchin, *History, Ethics, and Emergent Probability*, 2nd edn., Canada, The Lonergan Web Site, 1999, p. 59.

[54] See Melchin, 60. See also Andrew Beards, *Lonergan, Meaning and Method. Philosophical Essays*, New York, Bloomsbury Pub. Inc., 2016, p. 39.

There have been some works in the period 2002–2017 that have cited and been positive about *RSE*. But these have not attempted to implement its method.[55] Patrick Byrne recalls "the great contribution of *Randomness, Statistics and Emergence* ... to show how that work, and the work of Lonergan that it advanced, has implications beyond its explicit discussions to issues such as those raised by Karl Popper."[56] His paper, however, is in the mode of traditional philosophical comparison. But as mentioned in the Preamble,[57] it is a group problem, for the new method is not yet part of either the philosophic or scientific traditions.[58]

An exception is found in what Ken Melchin does in his book. In the introduction, he observes that:

> McShane's presentation is worked out in conversation with other positions in the field of the philosophy of science. And while my study is not intended as a contribution to the philosophy of science, some of the questions and issues dealt with by McShane provide helpful clues and a fuller exposition of the basic insights.[59]

Melchin's book goes on to challenge readers to reach a preliminary "grasp of grasping" elementary empirical probabilities; as well as a preliminary grasp of reaching probable judgement. Some of Robert Henman's research has helped reveal the need for generalized empirical method in neuroscience. [60] In "Randomness, Emergent Probability and Cosmopolis"[61]—as Melchin's work, also drawing on examples—James Duffy gives an introduction to some of the ideas developed in *RSE*. He goes on to discuss "emergent probability" within the historical context of "human affairs generally." [62] Duffy's paper climbs to a

---

[55] See, e.g., Patrick Daly, "An integral approach to health science and healthcare," *Theoretical Medical Bioethics*, vol. 38, 2017, pp. 15–40; Frank E. Budenholzer, *Zygon*, vol. 39, no. 2, 2004, pp. 339–356; Donna Teevan, "Albert Einstein and Bernard Lonergan on Empirical Method," *Zygon*, vol. 37, no. 4, 2002, pp. 873–890.

[56] Patrick Byrne, "Statistics as Science: Lonergan, McShane and Popper," *Journal of Macrodynamic Analysis*, vol. 3, 2003, p. 75.

[57] See Preamble, par. 5 on page ii.

[58] For readers familiar with Lonergan's writings, I note that implementation will be a future achievement that will be constitutive of the emergence of a "third stage of meaning" in history. Lonergan, *Method in Theology* (1972), pp. 93–99; CWL 14, 90–95.

[59] *History, Ethics, and Emergent Probability*, p. 7.

[60] Robert Henman, "Can brain scanning and imaging techniques contribute to a theory of thinking?" *Dialogues in Philosophy, Mental and Neuro Sciences*, vol. 6, no. 2, 2013, pp. 49–65.

[61] James Duffy, "El azar, la probabilidad emergente y la cosmopolis," *Revista de Filosofía* (Universidad Iberoamericana), *Debate Hermenéutica*, 135, Año 4, 2013, pp. 313–337. Translation by author.

[62] Eugene P. Odum, *Fundamentals of Ecology*, Philadelphia, Saunders, 1959, 26. Quoted in *RSE* (1970), 220; page 182 below.

discussion of Lonergan's discovery of "a third way," [63] the structure of "cosmopolis." He then raises questions about probabilities of emergence of schemes of recurrence in education, economics, and politics that will *not* "make life unlivable." [64] "Is it possible to change the probabilities of schemes of recurrence in education, economics and politics from a fraction $(p * q * r)$ to a sum $(p + q + r)$? How? Are we to ask philosophers and theologians to understand the basics of a healthy economy?"[65] Or again, in a more recent paper,[66] Duffy makes use of *RSE* as a resource. Among other things, that paper will help readers make beginnings toward a control of meaning in probability. For my part, what I have learned (and continue to learn) from *RSE* contributes to my ongoing preliminary efforts in implementing and writing about the need and possibility of the balanced empirical method.[67]

The paucity of scholarship that promotes or attempts to implement generalized empirical method is not unique to works that refer to *RSE*. This is not the place to attempt historical analysis. For present purposes, a sample text makes the need conspicuous by its absence. I draw your attention to a recent paper by Alexander Bird, a leading scholar known for his work on "scientific progress." The abstract of the article opens with a wonderful question. But in self-attentive mode, watch where Bird goes with the discussion.

> What constitutes scientific progress? This article considers and evaluates three competing answers to this question. These seek to understand scientific progress in terms of problem-solving, of truth-likeness/verisimilitude, and of knowledge, respectively. How does each fare, taking into consideration the fact that the history of science involves disruptive change, not merely the addition of new beliefs to old beliefs, and the fact that sometimes the history of such changes involves a sequence of theories, all of which are believed to be false,

---

[63] Duffy "El azar, la probabilidad emergente y la cosmopolis," p. 329. See also CWL 14, 8.

[64] Duffy, p. 336. "[P]hilosophers for at least two centuries, through doctrines on politics, economics, education, and through ever further doctrines, have been trying to remake man, and have done not a little to make human life unlivable." Lonergan, *Topics in Education,* Collected Works of Bernard Lonergan, vol. 10, Robert Doran and Frederick Crowe (eds.), University of Toronto Press, 1993, p. 232.

[65] Duffy, p. 336. See also Philip McShane (ed.), "Do You Want a Sane Global Economy?" *Divyadaan: Journal of Philosophy and Education*, vol. 21, no. 2, 2010. This volume was republished by *Journal of Macrodynamic Analysis* and is available online at https://journals.library.mun.ca/ojs/index.php/jmda/issue/view/134, (accessed December 1, 2020).

[66] James Duffy, Cecilia Moloney, and Terrance Quinn, "Assembling the Meaning of Probability," *Journal of Macrodynamic Analysis*, vol. 13, 2020, pp. 84–118.

[67] See Terrance Quinn, Books, Articles, terrancequinn.com.

even by scientific realists? The three answers are also evaluated with regard to how they assess certain real and hypothetical scientific changes. Also considered are the three views of the goal of science implicit in the three answers. The view that the goal of science is knowledge, and that progress is constituted by the accumulation of knowledge is argued to be preferable to its competitors.[68]

After asking the question 'What is scientific progress?' Bird's focus goes to "three competing answers." In the article, those "answers" are discussed in terms of various "philosophic conceptions" with no concrete linkages given to the author's experience in scientific progress. Why is one conception "preferable to its competitors"?[69] Is it a matter of having a better argument? Whatever *progress in philosophy of science* might be,[70] the paper that nominally is about scientific progress is written in a linguistico-analytic mode, remote to instances of actual scientific progress. Again, though, this is not to single out Bird's paper. The article is part of a tradition of speculative dialogue about science that, as mentioned in the Preamble, goes back at least three centuries.

3.   Reading *RSE* in 2020: some advances made in *RSE*

3.1   Goals, method, structure, and content

I begin by noting a main goal of *RSE*, intended by the author:

> **The book might well have been subtitled *Towards an adequate Weltanschauung.*** Such a subtitle may appear somewhat extravagant since the topics dealt with seem to be special questions in the philosophy of science. As I have argued elsewhere,[71] however, in so far as one is a serious thinker, claiming an adequate viewpoint, a central element in that viewpoint is one's thought on the relationship of chemistry to botany. Without that thought one lacks a basic component for the conception of world process.[72] The present work deals with the central element and the heuristic conception of world process. **It tries to lead the reader towards an adequate *Weltanschauung* through a dialectic of personal performance. The world view that should emerge—in the reading subject—is the world view of *Critical Existentialism*, a view which originated with Bernard Lonergan. Critical Existentialism contrasts with other forms of existentialism in that it seeks to mediate the subject's authen-**

---

[68] "Scientific Progress," in *The Oxford Handbook of Philosophy of Science,* p. 544.

[69] See note 68.

[70] See note 37 .

[71] Philip McShane, *Music That Is Soundless*, Dublin, Milltown Institute, 1969. Cited in *RSE* (1970), ix, note 3; page lxv below.

[72] See also note 128.

**ticity through a self-appreciation of metascientific dimensions.
... which would be adequate to our times.**[73]

McShane then describes the structure and content of the book as "an attempt to do four things at once."[74]

> [The first] is an attempt at inaugurating dialogue between various schools of philosophy. ... [The second] is **an attempt to orientate philosophy of science away from general considerations towards a reflection on the detailed content of science and on the details of procedure in scientific investigation.** ... [The third] offers a more detailed account of various points treated in the works of Lonergan, particular of course the principle of emergent probability, but also more precise indications, such as a possible meaning of the change from the first to the second edition of *Insight* on the question of convergence in probability sequences. ... [Fourthly] there is the central positive contribution of the thesis. It is an attempt to establish on a wider basis of contemporary mathematics and science the position of B. Lonergan on the nature of randomness, statistics, and emergence.[75]

You might notice that the second point is about method. *RSE* is an attempt to be "empirical and scientific in a generalized sense."[76] I would not presume to know McShane's meanings for this terminology (1970ff), nor Lonergan's (1957ff). But it is safe to say that "generalized" does not mean "discussing in general terms" (as in the analytic tradition). On the contrary, "scientific in a generalized sense" is, in the first place, "scientific." One is to advert to all relevant data. Familiar sources of data are not excluded. But data (experience) in science includes one's questions, insights, and formulations. And so, "empirical and scientific in a generalized sense"[77] is *adequate empirical method*, or simply Empirical Method.[78] In the second point above, then, we see McShane departing from conventional philosophy of science. For, in generalized empirical method the task for the philosopher of science is not one of "philosophy from afar." The task is, instead, that of **"transforming and correctly orientating science from within,**

---

[73] *RSE* (1970), ix; page lxiv below. Boldface added here and in quotations below.

[74] *RSE* (1970), vii; page lxiii below.

[75] *RSE* (1970), vii-viii; pages lxiii–lxiv below.

[76] *RSE* (1970), 253; page 209 below. See section 1, par. 1, and ff.

[77] *RSE* (1970), 253; page 209 below.

[78] Frederick Lawrence, "The Ethics of Authenticity and the Human Good, in Honour of Michael Vertin, an Authentic Colleague," in John J. Liptay Jr. and David S. Liptay (eds.), *The Importance of Insight: Essays in Honour of Michael Vertin*, University of Toronto Press, 2007, p. 131.

and [s/]he can do so adequately only in so far as [s/]he includes as data for philosophic reflection the details of contemporary science."[79] Consequently,

> [t]he argument [in *RSE*] ... runs on two main levels. Firstly, it moves cumulatively toward particular sets of results. Secondly it moves, more implicitly but still cumulatively, towards a justification of the method by which the results were reached.... [I]t is as well to note here that there is presupposed throughout a view on objectivity which is of a piece with the philosophic method. It is of a piece with the philosophic method, which is one of attention to oneself in the performance of knowing, because it can only be established thematically through the subject's attention to his or her own performance in knowing. Briefly, **the basic assumption is that the real is what is to be reached by correct understanding. That assumption cannot be intelligently challenged without being performatively contradicted in the process of challenging: it is the basic assumption of all intelligent dialogue.**[80]

McShane's goal was ambitious. One can be working in a mode of "self-attention" but be centuries behind the times. There is creative learning. There is also creative progress that advances frontlines. In *RSE*, the goal includes being "adequate to our times."[81] And so the challenge that McShane took on included "self-attention in the process of doing the relevant [modern] sciences"[82] that had, in fact, been making progress (and continue to do so). His concluding description of the method and results are as follows:

> The present effort goes beyond the analytic effort in giving closer attention to all that is contemporarily known through common sense and science as it emerges in the mind of the knower. The close attention involved was precisely specified as self-attention, and the result of that self-attention is an appreciation of structured anticipations of what is to be known, ranging from the simple triple structure of every scientific effort to an intricate contemporary anticipation of an understanding of world process.[83]

Did McShane succeed in his main goal? Was the method valid? Were details and results obtained correct? Is the book *RSE* worthwhile? My answer to these questions is "Yes." I have learned much from *RSE* and continue to learn.

---

[79] *RSE* (1970), vii; page lxiii below. See also note 4.
[80] *RSE* (1970), 13; page 10 below.
[81] See the quotation at note 73 above, on pages xiii–xiv.
[82] *RSE* (1970), 257; page 212 below.
[83] *RSE* (1970), 260; page 215 below.

Moreover, I am confident that it can be a help toward progress, globally. I cannot discuss all of the important details. I have found that each page of the book has riches to be plumbed and to which, over the years, I have returned. In sections 3.2 and 3.3, I invite your attention to what I have found to be some key results in *RSE*, "marquee moments" as it were.

Because the approach advocated by *RSE* remains novel, the following point is worth repeating: If you read the book in the traditional manner of philosophy of science (e.g., with a focus on language and logic), you will not be reading the book in the way intended by the author. More importantly, you will be missing out on personal development invited.[84] If, however, you choose to "enter the show," I am confident that you will find it increasingly (self-) evident that "the argument genuinely accumulates and [that results can be] appreciated within the adequate perspective, which is one of **scientific self-attention on the part of the reader.**"[85] Borrowing a term from the mathematics of functions, you will find that you, the reader, are "the (self-) *argument*" that "accumulates." A cautionary note also can help you prepare for the challenge: "[I]t is not an easy matter to identify precisely what one is understanding in a diagram, whether the diagram is one of a geometrical problem or of biochemical processes."[86] In my experience, the same comment applies to all essential aspects of generalized empirical method in modern science and philosophy of science.

3.2 Abstraction, randomness, statistics, and probabilities

A shortlist of important results in *RSE* is provided by McShane himself: there are eight problems each in List A: Statistics and List B: Emergence, respectively.[87] There are corresponding answers in two follow-up lists, also called A and B.[88] You might find it helpful to have a look at those lists immediately. This will help give you a sense of the scope and subtlety of the work. But as did McShane, I provide the lists not to provide a "summary"[89] but "to raise the questions which will be considered in the concluding chapter and, in this way, to make possible a change in the focus of attention in that reading."[90]

Chapter 2 of *RSE* includes reflection on attempts to mathematically define *randomness*. I have found the chapter a helpful stand-alone contribution to foundations of mathematics.[91] For present purposes, however, I hold off from

---

[84] *RSE* (1970), 11, lines 11–18; page 8 below.

[85] *RSE* (1970), 46–47; page 37 below.

[86] *RSE* (1970), 199; page 164 below.

[87] *RSE* (1970), 4–5; pages 3–4 below.

[88] *RSE* (1970), 9–10; page 7 below.

[89] See note 17.

[90] *RSE* (1970), 10; page 8 below.

[91] It is an aspect of the problem. See also Philip McShane, "The Foundations of Mathematics," *The Modern Schoolman*, vol. 40, 1963, pp. 373–387.

discussing much content from that chapter. I leave such efforts to another time, and to those who will contribute to future recyclings and implementations of *RSE*.[92] However, I do invite your attention to an observation reached in the chapter, namely, that "one must ultimately acknowledge that the object [random variation] of this type of investigation is strictly empirical and that it requires to be specified as such, as something-to-be-understood."[93] That observation helps nudge us along toward the main goal of the book. There is one more thing from chapter 2 to which I also call your attention, a result which does not regard randomness. McShane provides a "simple illustration as paradigm"[94] of what is sometimes called *Aristotelian abstraction*.

McShane asks readers to re-enact discovery of Boyle's law $PV = C$; to discern elements of one's discovery; and so, to identify the enriching aspect of abstraction. One needs to be (or become) familiar with experimental arrangements and to have access to data (historical, or otherwise). Being able to (self-) identify abstraction (in this or some similar instance[95]) will be an increment in growth toward becoming luminous in explanation and description.[96]

McShane provides the following description of the exercise:

That law [$PV = C$], far from being a relation between impoverished replicas, can be seen to result from a **correlation of correlations of correlations of data. The triple 'correlation' is deliberate.** The key to the movement towards the law lies in measurement, and the initial correlation results from the selection and use of a basis of measurement. There is then the process of correlation involved in providing suitable measuring apparatus. There is next the sequence of correlations of measuring apparatus with measured which yields in our case a list of volumes or a list of pressures. The lists are at this stage

---

[92] I am, in particular, anticipating the emergence of functional collaboration. See section 4.3.

[93] *RSE* (1970), 31; page 24 below.

[94] *RSE* (1970), 47; page 37 below.

[95] For instance, one might re-enact Galileo's discovery of the law of falling bodies. In that case, however, one needs to join Galileo in inventing a way to measure "time."

[96] McShane adverts to this issue later in the book. The "distinction between explanation and description would seem to be more than a terminological convenience, for the transition from prescientific description to scientific explanation involves the discontinuity of a switch from relatedness of things to us to correlation of things among themselves. It is the discontinuous switch from heat as felt to heat as defined by the correlations implicit in thermodynamic equations, or from colours as seen to wave and photon theories. The relation between the two brings up the problem of correspondence rules, but at all events it is clear that without the prescientific description there can be neither transition nor correspondence rules." *RSE* (1970), 172; page 143 below.

materially associated. Finally, the central insight giving rise to the law, the correlation, is that which associates the two lists—abstracting concomitantly from errors and random differences—to arrive at the relation of inverse proportion. The scientist will have no difficulty in recognizing the description of the procedure, which is typical of the emergence of any scientific law. That scientific law is abstract, general, making no mention of particular cases or particular values.[97]

RSE invites readers to sort out various notions in statistics and probability that, to this day, remain unresolved in the literature. For instance, the book will help you make progress in (reaching preliminary control of meaning in) distinguishing *reasonable betting* and *chance*. One also is invited to (self-) observe the *complementarity* of *classical method* and *statistical method*.[98] It is to be noted

> that the word 'complementarity' as used here has nothing to do with Bohr's principle of complementarity. Bohr's principle is linked with physics and has its origin in the fact that pairs of conjugate variables are normally associated, one with periodic, one with non-periodic properties of microentities. There is room for complementarity in our sense in all the sciences: it is a complementarity of statistical and classical explanations which aids towards the goal of complete explanation.[99]

The complementarity in methods is concrete and is found at all levels of scientific inquiry. An emphasis of what can be named *classical method* is to describe and define events. An emphasis of what can be named *statistical method* is to determine whether or not, and to what extent, events (as defined or described) occur.

Clarification becomes possible regarding what at first glance might seem to be (but is not) a minor technical matter.

> [T]he continuous functions of classical theory refer to a continuum of events, the theoretical continuity of processes such as a planetary orbit. But in statistical theory the continuity of functions refers, not to a continuity of process but to the continuity of the ideal norm from which the observed values diverge at random.[100]

---

[97] *RSE* (1970), 47; page 37 below. For some help with this, see e.g., Robert Henman, "Boyle's J-tube experiments," pp. 3–4, https://bentonfuturology.com/wp-content/uploads/2020/09/Boyles-Law-of-Pressures-and-Volumes-by-Robert-Henman.pdf, (accessed December 1, 2020).

[98] See also Chapter 4 "The Complementarity of Classical and Statistical Investigations," CWL 3, 126–162.

[99] *RSE* (1970), 83; page 68 below.

[100] *RSE* (1970), 55; page 43 below.

Once intussuscepted within contemporary contexts, this will provide new control of meaning in work that, e.g., allegedly contrasts "stochastic differential equations" and "deterministic differential equations." But in many applications (e.g., population dynamics) solutions of systems of ordinary differential questions are not of 'deterministic' processes but merely provide central means about which actual population dynamics vary randomly. In practice, whether or not results are "classical" or "statistical" is not determined by mathematical structure but by concrete referents.

With examples, *RSE* invites one to notice and self-notice that the word 'probability' has different meanings. It is one name but there are two distinct *operations* to consider. These core distinctions mainly have neither been adverted to nor accurately described. But by engaging in exercises (in self-attention), readers will, in particular, be helped toward becoming luminous in the fact that "statistical science is not a science of the singular"[101] and that *'degree of belief'* and other numerical constructs attached to judgement are mere metaphors. By adverting to and discerning shifts in mental stance,[102] one can precisely distinguish "v-probability"[103] (quality of an is it so? answer) and "f-probability"[104] (an answer to a 'what is it? question, where the what is it? poise is to discover an "ideal relative frequency").[105] Focusing on source insights, the "vexed question

---

[101] *RSE* (1970), 55; page 43 below.

[102] *RSE* (1970), 135, 136, 138, 141, 143, 144; pages 112, 114, 116–117, 118, 119, below.

[103] *RSE* (1970), 137–143; pages 114–118 below.

[104] *RSE* (1970), 137–143; pages 114–118 below.

[105] The v- and f-probabilities are verifiable in experience. In the literature, there is "mathematical probability," which is a ratio, proportion or measure that need have no empirical referents. (See also the paragraph following note 111.) What sometimes is called an "experimental probability" is merely a relative actual frequency. What McShane calls v- and f-probabilities also are to be distinguished from the philosophical notions of 'objectivist' and 'subjectivist' probabilities. Basic confusions are present in modern textbooks at all levels. For example, see the first-year university textbook, Richard De Veaux et al., *Stats, Data and Models, Third Canadian Edition*, North York, ON, Pearson, 2019. In this book, *probability* is defined to be "a number between 0 and 1 that reports the likelihood of an event's occurrence" (356), in that way inadvertently combining v-probability with f-probability and at the same time mistakenly attributing f-probability to singular events; *empirical probability* is defined to be "the probability that comes from the long-run relative frequency of an event's occurrence" (356), in that way attempting to be empirical but avoiding the fundamental problem of defining convergence (see note 115) and not adverting to the fact that even in large sample sets random differences occur; *theoretical probability* is defined to be "when probability comes from a mathematical model and not from observation" (358), revealing a lack of control of meaning in distinctions between mathematical and empirical probability (see note 117 and text ff); and *personal probability* is defined to be one's "personal assessment of your

of convergence"[106] also can be handled. And so, one is helped toward reaching understandings of *empirical probability* and *statistical state* [107] that are not philosophical in traditional terms but are verifiable in scientific practice.

Identification of the two types of probability also allows for precision in resolving various common errors. For instance, in the philosophical literature as well as in the (ongoing) tradition of (undergraduate and graduate) textbooks on probability and statistics, "unit probability" is identified with (a notion of) "certainty" not found in human experience. Following McShane's lead—adverting to experience in scientific practice—you will reach a preliminary "control of meaning" in the fact that

> unit [f-]probability ... has nothing to do with the acceptability of a hypothesis, proposition, or system, nor is it a quality of an is-answer. Unit probability can be a hypothesis for a certain class of events, being verified in so far as the divergence of the frequency of occurrence of those events from unity is random. As any hypothesis, it is the answer to a what-question. It is the peculiar limiting case of the fraction which gives an ideal frequency, and so the non-systematic divergence from it can only occur by actual frequency falling below it, but otherwise there is nothing unusual about it. The hypothesis, $p = 1$, f-probability, can be verified, and the verification can be more or less probable, v-probability.[108]

Growth in control of meaning in the two types of probability also provides for "clarification of the nature of verification in general."[109] However, as with the mathematical content of chapter 2 of *RSE*, I leave that topic for future collaborations.

Chapter 8, "Foundations of Statistics," continues in the work of bringing out the possibility of "orientating science from within."[110] McShane points out that the chapter

---

chances of getting" (359) a particular outcome, "expressing your uncertainty about the outcome" (359), thus describing but not identifying experiential and cognitional sources of v-probability. The same confusions are carried into discussions regarding mutually exclusive events, confidence intervals (e.g., "95% confidence" (479)), as well as into upper-division and graduate level texts. See, e.g., Peter Olofsson and Mikael Andersson, *Probability, Statistics and Stochastic Processes*, Hoboken, NJ, Wiley & Sons, Inc., 2012, sec. 1.1. As mentioned in the Preamble, these issues will become self-evident if one self-attentively works through examples provided in *RSE*.

[106] *RSE* (1970), 149; page 123 below.
[107] *RSE* (1970), 149; page 123 below; CWL 3, 81–89, secs. 4.3–4.4.
[108] *RSE* (1970), 143; page 118 below.
[109] *RSE* (1970), 131; page 109 below.
[110] *RSE* (1970), vii; page lxiii below.

may be regarded as an aside upon which the following chapters do not depend. **The aside is, however, worthwhile** because there are implicit here basic difficulties of a methodological nature, the discussion of which throws light not only on the question in hand but on the general problem of heuristics, of the passage from description to explanation, from nominal to essential definition.[111]

One of the results obtained in chapter 8 corrects a view still held by some (or at least about which there is not yet clear consensus), namely, that probability theory is an area of mathematics. At the same time, there remains a practical need for "frequency theories." Recall, however, that in *RSE* the focus is on "preformulation."[112] And so, "[o]ur problem … is not to solve any mathematical difficulties, but to expose what exactly is involved in these developments, how they are related to the empirical, where there are weaknesses in the discussion."[113] But there remain weaknesses in the "usual frequency theory of probability."[114] The central problem is "to give precise and acceptable 'meaning' to the notion of 'clustering of frequencies' which, as we have seen, is at the very root of the interpretation of randomness."[115]

Heuristic formulas are sometimes used to express the idea that if sample sets are sufficiently large, and if we have sufficiently large numbers, m say, of trials, then averages of relative actual frequencies obtained from m large samples "should in some way converge" to the correct probability p for the event in question. That heuristics is sometimes expressed by an ill-defined limit form:

$$\lim_{m \to \infty} \frac{1}{m} \Sigma_1^m \left(\lim_{n \to \infty} \frac{S_{nm}}{n} \right) = p. \text{ [116]}$$

As McShane points out, "[a]ll this inaccuracy and looseness will undoubtedly cause the mathematician to groan. What is called for is a more accurate formulation, one indeed which will both avail itself of and contribute to the development of convergence theory."[117]

---

[111] *RSE* (1970), 150; page 123 below. See also note 96.

[112] *RSE* (1970), 151, 155, 157; pages 125, 128, 130 below.

[113] *RSE* (1970), 151; page 124 below.

[114] *RSE* (1970), 151; page 125 below.

[115] Loève, *Probability Theory*, 14. Quoted in *RSE* (1970), p. 152, n. 18; page 126 below.

[116] *RSE* (1970), 154; page 127 below. Similar heuristic formulas are provided in contemporary textbooks. For instance: "In mathematical notation, if we consider n repetitions of the experiment and if S_n of these gave our outcome, then the relative frequency would be $f_n = \frac{S_n}{n}$, and we might say that the probability equals $\lim_{n \to \infty} f_n$." De Veaux et al, *Stats, Data and Models*, Sec. 1.1.

[117] *RSE* (1970), 154; page 127 below.

Sorting through details of main views on the matter, McShane invites us to discern subtle shifts in meaning that reveal advances as well as various underlying confusions. As modern science confirms, he anticipates "the possibility of development in probability theory through the distinction of various types of empirical randomness."[118] It becomes (yes, self-) evident that "[t]he theory of probability is not just 'like all mathematics': it is more closely concerned with the concrete than classical mathematics."[119] And the mathematical Laws of Large Numbers do not close the gap. Further precisions follow, including the following:

> As the parallel postulate is to nominal definition of the plane surface, so will any attempt to formulate an intelligibility of empirical prob-ability be to its common matter. The common matter in the latter case is such as to elude neat specification, as our earlier discussion of randomness showed. ... It is not then a matter of working out mathematically a series of manageable definitions of randomness etc.—though this too is involved—but of taking into account the contrariness of empirical randomness in its various types ... **Our basic point throughout is that advance in probability theory requires the transformation of the central insight**[120] **according to the particular conditions of randomness to which the theory is relevant.**[121]

3.3 Things, emergence, schemes, and all

In chapter 9 of *RSE*, McShane details various attempts in the literature to account for the fact that, in cellular processes, there is physics *and* chemistry. For example, Rashevsky's approach is to model "the cell" with systems of differential equations. This is worth looking at here. For while there have been advances in methods in modern systems biology (genomics, proteomics, cellular automata, and so on), main aspects of the Rashevsky paradigm survive.[122] Important results

---

[118] *RSE* (1970), 157; page 129 below.

[119] *RSE* (1970), 164; page 136 below.

[120] Discussed and noted in chapter 2: "In chapter eight we will return to these foundation questions and we will see that there seems to be room for the specification of subtypes of empirical results and for the emergence of corresponding theories of probability." *RSE* (1970), 31; page 24 below.

[121] *RSE* (1970), 166; page 137 below; *RSE* (1970), 169; page 139 below.

[122] See, for example, current trends in the *Journal of Mathematical Biology*, https://www.springer.com/journal/285, (accessed December 1, 2020). Two recent papers that illustrate the orientation are Yifei Lim, Peter van Heijster, and Matthew J. Simpson, "Travelling wave solutions in a negative nonlinear diffusion–reaction model," *Journal of Mathematical Biology*, vol. 81, 2020, pp. 1495–1522; and Maria Eckardt, Kevin J. Painter, Christina Surulescu, and Anna Zhigun, "Nonlocal and local models for taxis in

are possible, such as obtaining upper and lower limits on cell membrane dynamics. However, as McShane points out, there are limitations to the approach. For the positive results so obtained regard a small number of variables selected from an abstractly defined system of equations and are computed by averaging. By definition, such results do not pertain to any particular cell but instead help reveal possible ideal averages for aggregates of aggregates of aggregates (cells, chemical events, physical events). In fact, as defined in the approach, a system of differential equations for the physics and chemistry of a cell is non-verifiable. Hypothetical boundary conditions for all chemical and physical quantities of a cell are not available to observation. And even if one were able to accomplish a super-computer estimate of consequent physical and chemical events, this would be another aggregate of aggregates and would not be explanatory of concretely verifiable cellular functions. So, while not to suggest it is not useful in some respects, evidently and self-evidently, neither in its heuristics nor in its application does Rashevsky's approach provide a solution to the biological problem of explaining an individual one-celled organism, let alone a multi-cellular organism. And we will do well to recall that in scientific understanding, "we are dealing neither with a superman's grasp nor with the knowledge of a Laplace's demon, but with ordinary human intelligence."[123]

McShane also draws attention to Waddington's good suggestion, namely, that "[t]he secret to their [cells'] performance in this way is architecture, or to use the Aristotelian term, form."[124] However, in modern science, "it is not sufficient merely to recall an outworn terminology of matter and form: the Aristotelian couplet must be properly and contemporarily conceived."[125] **If one follows McShane here and works through the "exercises," it will become (self-) evident that "[i]f one does not venture into biochemistry and biophysics, one remains on the descriptive level in biology."**[126] He then begins to home in on a key result of *RSE*:

> Another way of expressing the relation of proportionate matter to form is to note that different types of chemical aggregates give different biological types, and this on the various levels of cell, organ, organism. It would take us too far afield to go into details here but it is worth noting that a general account of genera and species on various levels could be developed on this basis: a series of definite coincidental

---

cell migration: a rigorous limit procedure," *Journal of Mathematical Biology*, vol. 81, 2020, pp. 1251–1298.

[123] *RSE* (1970), 181; page 149 below.

[124] C. H. Waddington, *The Nature of Life*, London, Allen & Unwin, 1963, p. 21. Cited in *RSE* (1970), 190, note 56; page 157 below.

[125] *RSE* (1970), 190; page 157 below.

[126] *RSE* (1970), 201; page 166 below.

aggregates of processes, say on the chemical level, would be in one-one correspondence with a series of biological laws.[127]

But details of the solution-heuristics are given later in the book:

[O]ne can … approach the problem of understanding the amoeba against the background of physics and chemistry that we discussed in chapter nine. Most evidently, this advance into physics and chemistry represents a movement from description to scientific explanation. But what is important in the present context is the cognitional difference which corresponds to a transition from sensible presentation to symbolic representation. Clear illustrations of such transitions are most easily found on the level of mathematics. Thus a cartwheel or the sketch of a round plane curve can lead to the question, Why is this round?, and to the definition of a circle. One is dealing here with sensible presentations. But the image can be less adequate than this, as when one draws two straight lines and supposes them to be parallel. **The important instance of presentation for us, however, is the symbolic presentation** in which the symbols stand for whatever one assigns them to stand for. The simplest illustration of this type of image is the drawn or imagined straight line which can neither be drawn nor imagined without breadth, but which stands for or is thought of as without breadth. **But the illustration most relevant to our purpose is that which occurs in discussions of the amoeba against the background of physics and chemistry. Here, instead of operating with the sensibly-presented amoeba, the scientist operates in the context of symbolic images of the chemical and physical processes.** Examples of the procedure were in fact given in chapter nine. The process, indeed, is familiar to the scientist in this field; he makes use of it in textbooks, in teaching, in research; but he does not make explicit, as we try to do here, what precisely is going on. Instead, then, of the microscopically-observed amoeba he has the image of the coincidental aggregate of chemical or cytological reactions, and as we saw at some length in chapter nine, the biological relations provide him with a systematic understanding of these coincidental aggregates.

So, we come to see that the basic analogy, form is to matter as understanding is to presentation, leads to an answer to our initial question regarding the relation of the biological to the biochemical. In cognitional terms, the symbolically-presented

---

biochemical level is to biological understanding as matter to form.[128]

Let's now go back to McShane's windup of chapter 9. There is a need to integrate results about understanding the amoeba with results obtained on randomness and emergence:

> Our conclusion at the end of this long section ... is fourfold.
>
> First, we have succeeded in specifying further the meaning of randomness and in increasing the evidence for its objectivity. ...
>
> Secondly, we have associated the problem of objective randomness with the problem of distinct levels of investigation, distinct sciences. ...
>
> Thirdly, we have clarified various aspects of the problem of emergence. The possibility of emergence was seen to be intimately linked with the occurrence of randomness, of the non-systematic, at a given level. ...
>
> Lastly, the first three conclusions must be qualified by a point made repeatedly already. The relevant meaning of randomness, of emergence, and of their objectivity, will be in the reader's mind only in so far as the reader has been operating in the mode of scientific methodology.[129]

Chapter 10 of *RSE* looks to *recurrence-schemes* (*schemes of recurrence*), emergence of recurrence-schemes, and *schedules of probability*.

What are *recurrence-schemes*? The name is not to be confused with similar terminology in contemporary computational science. Naming aside, examples in world process include planetary orbits; "biochemical cycles" internal to an organism; life cycles of organisms; cyclic patterns in multi-species predator-prey-alimentation dynamics; oxygen, nitrogen, carbon, and water cycles, and other inorganic and organic "circular" sequences of mutual dependence in global ecosystems.

McShane describes a recurrence-scheme as follows: "a systematic arrangement which includes concrete determinations and the systematic arrangement is one which ensures repetition."[130] As defined by Lonergan, "positive conditions for an event might coil around in a circle. In that case, a series of events A, B, C ... would be so related that the fulfillment of the conditions for each would be the occurrence of the others. Schematically, then, the scheme might be represented by the series of conditionals: if A occurs then B will occur; if B occurs then C will

---

[128] *RSE* (1970), 251; page 207 below. This helps concretize the advanced densely expressed heuristics given in CWL 3, 489.

[129] *RSE* (1970), 203–204; pages 168–169 below.

[130] *RSE* (1970), 219; page 182 below.

occur; ... *A* will recur. Such a circular arrangement may involve any number of terms, the possibility of alternative routes, and in general any degree of complexity."[131] An instance of greater complexity needs to be noted: a "scheme might consist of a set of almost complete circular arrangements of which none could function alone yet all would function if conjoined in an interdependent combination."[132]

McShane points out that "as is evident from most of the examples, the cyclic or circular nature of the recurrence [in a recurrence-scheme] is not physical but metaphorical."[133] There is, for instance, the TCA or Krebs cycle. What is empirically verifiable is not a "cycle of reactions in cells" per se but, rather, a cyclic pattern of mutually related biochemical capacities-to-perform of an organism.[134] Moreover, that realization in the scientist emerges from patterns of collaboration in history which also are schemes of recurrence in "human affairs generally."[135]

Schemes of recurrence are not conceptual constructs. They are concretely possible. There are periods of time during which a scheme of recurrence is a "going concern."[136] In evolutionary studies, part of what is of interest is that new recurrence-schemes emerge. Some can be relatively enduring, while others are not and become extinct. There is, then, an emerging challenge in modern science, to compare and count schemes of recurrence. In that way, relative actual frequencies of functioning schemes of recurrence will provide data needed for discovering and verifying empirical probabilities of functioning schemes of recurrence. In a similar way, there is also the need and possibility for working out *empirical probabilities of emergence* of recurrence-schemes, probabilities of stability (or not), of survival (or not),[137] and so on.

However, the way in which to concretely determine relative actual frequencies and empirical probabilities of schemes of recurrence is not obvious. That there will be subtleties involved can be glimpsed by looking to some familiar

---

[131] CWL 3, 141.

[132] CWL 3, 141.

[133] *RSE* (1970), 227; page 188 below.

[134] This is not a philosophical claim. One needs to work through lab results in the mode of generalized empirical method. See note 129. Contemporary in vivo verification is often obtained by introducing ^{13}C tracers. Nuclear magnetic resonance reveals approximate concentrations of biophysical events that within errors bounds are compatible with the biochemical equations of the TCA cycle. See, e.g., Clinton M. Hasenour, Mohsin Rahim, Jamey D. Young, "In Vivo Estimates of Liver Metabolic Flux Assessed by ^{13}C-Propionate and ^{13}C-Lactate Are Impacted by Tracer Recycling and Equilibrium Assumptions," *Cell Reports*, vol. 32, 2020, https://doi.org/10.1016/j.celrep.2020.107986, (accessed December 1, 2020).

[135] *RSE* (1970), 220; page 182 below.

[136] *RSE* (1970), 221; page 183 below.

[137] *RSE* (1970), 231; pages 191–192 below. See section 4.1.

schemes of recurrence. McShane begins with the example of the "orbiting of the earth round the sun."[138]

> Under the unfavourable circumstances mentioned the members of the set may be said to be practically unrelated, and so the probability of occurrence of the whole set can best be represented by something like the product of the probabilities of the members. But if the circumstances change—in our simple parallel, something akin to a pressure-reduction in the gas—then the set of initial conditions are no longer unrelated. If the earth comes into any one of the set of conditions involving the suitable position and velocity for the particular elliptic orbit, all other members of the set defined by that elliptic orbit will occur. **Since therefore, if any of the set occurs, all occur, the change of circumstances would seem to have effected something like a jump in probability from the product to the sum of the particular probabilities.** Needless to say, our example is extremely simple and so not representative of concrete problems of calculating probabilities in evolution theory. But at least one can appreciate clearly in this instance what will be less clear and precise in more complex instances, the change in the probability of emergence on the fulfilment of prior conditions or the provision of adequate environment. The notion is very general and heuristic and the possibility of filling out its details remote.[139]

Subsequent to emergence of a recurrence-scheme, there are *probabilities of survival*. These will be of special interest at a time in history when modes of human living have been increasingly undermining global ecosystems and human cultures. Regarding *probability of survival*, McShane again begins by looking to the simple example of the earth orbiting the sun: "Thus, in so far as such orbital schemes occur, their continuance in existence is conditioned by a variety of internal and external factors, and so for given circumstances probabilities of survival might be estimated."[140]

However,

> [a]t the other end of the scale of size a more complex example of the same thing could be provided by a discussion of the occurrence and survival of Krebs cycles. Again there is the need for the realization of the prior suitable biochemical conditions and the presence of the required components before the cycle becomes concretely possible, and once these conditions are realized the probabilities of the

---

[138] *RSE* (1970), 230; page 191 below.
[139] *RSE* (1970), 230–231; page 191 below. See also section 4.1.
[140] *RSE* (1970), 231; page 191 below.

component-reactions in the cycle tend to complement one another additively. Here, too, there is the problem of the probability of survival of such schemes which leads to a consideration of stability of environments, etc.[141]

Note McShane's consistent focus on "pre-formulational"[142] aspects of the problem. Specifics are to be determined by empirical investigators. Continuing in that mode, and with the distinction in mind between empirical probability and "reasonable betting," we can ask what "empirical probability of survival of a scheme of recurrence" might look like. At first glance, it may seem to be best left to v-probability. However, the survival of a functioning scheme of recurrence depends on the *nonoccurrence* of events that would disrupt the scheme.[143] And so (with "environment"[144] determined by circumstances), an empirical probability of such nonoccurrence will be something like 1 – (probability of (their) occurrence). In other words, the empirical probability of survival of a scheme of recurrence will (on subtraction from unity) be determined by relative actual frequencies and empirical probabilities of events that are found to in fact disrupt the scheme of recurrence. (For example, the survival of global ecosystems is adversely affected by increasing concentrations of carbon dioxide; and in a global pandemic, investigators in epidemiology look to incidence rates.[145])

Adding still more to the heuristics, McShane introduces *schedules of probability.*

Before continuing our treatment of the probability of schemes we had best clear up a point already referred to: that our application of probability to the evolutionary process is not in the same category as reasonable betting. When we speak of the probability of schemes or conditioned series of schemes we have in mind the strictly statistical notion of probability.

…

As we shall see, our application of probability does indeed give some understanding of the universe as a whole, but that application is neither a type of subjective confidence through reasonable betting nor a matter of considering aggregates of complete evolutionary processes. It is a much more prosaic affair, closely related at all stages to the empirical. As an illustration of what we have in mind we may consider the distribution of what C. S. Coon would consider the five major

---

[141] *RSE* (1970), 231; page 192 below.

[142] *RSE* (1970), 151, 155, 157; pages 125, 128, 130 below.

[143] CWL 3, 144.

[144] See note 141.

[145] Miquel Porta, *A Dictionary of Epidemiology*, 6th edn., Oxford University Press, 2014 (print), 2016 (online), doi:10.1093/acref/9780199976720.001.0001, (accessed December 1, 2020), incidence, rate.

types, Australoids, Mongoloids, Caucasoids, Capoids and Congoids, on the surface of the earth at any time.[146] Coon's maps for the pleistocene and early post-pleistocene concretely illustrate our point. Clearly, the pleistocene distribution gives a clue to the later distribution: scientifically speaking, on the basis of the actual distribution for the pleistocene and other relevant factors, a schedule of probabilities could be elaborated for the further stage. This schedule deals with a set of probable next stages. Again, from data on the next stage a similar schedule for a further stage may be elaborated, and so on, through successive schedules of probability for conveniently selected stages.[147]

The heuristics sketched for schemes of recurrence are based on established results and are drawn from a range of areas in modern science. However, schemes of recurrence were not (and still are not) featured in contemporary evolutionary studies. As I bring out in section 4.2 (below), to include schemes of recurrence in evolutionary studies will require major advances in scientific understanding, in method and in heuristics. No doubt, the "introduction of the scheme as unit would complicate considerably the statistics. ... Still, difficulty is not a criterion of scientific unacceptability."[148]

A high point of the book is in chapter 11, where McShane gathers results and looks to the possibility of an integral heuristics of world process.[149] However, he pauses to ask about Fisher's conjecture wherein some kind of multiple Markov matrix might give the required picture of history. In the model, matrix elements are time-step transition probabilities in an "m − space," where m is the number of independent variables, excluding time. McShane draws attention to what, for readers who have come that far in the book, are evident flaws in Fisher's model. It is

> a highly abstract affair, and one moreover which is rooted in Newtonian Mechanics. No doubt this approach may be of value in certain detailed investigations of economics,[150] but as a candidate for a general methodological account of evolution it is a non-starter. In so

---

[146] G. S. Coon, *The Origin of Races,* London, Knopf, 1963, "especially the maps on [page] 658." Cited in *RSE* (1970), p. 236, n. 10; page 196 below.

[147] *RSE* (1970), 235–236; pages 195–196 below. The results are dated but the principle remains. See, e.g., http://www.bbc.com/earth/story/20150929-why-are-we-the-only-human-species-still-alive, (accessed December 1, 2020). See also the examples given above, the emergence and co-existence of different types of plants involved in distinct but in many ways "overlapping" ecological schemes of recurrence.

[148] *RSE* (1970), 238–239; page 197 below.

[149] See note 73 above on pages xiii–xiv.

[150] See section 3.3, (g).

far as it does propose a sequence of, say, n schedules of probabilities through time it bears some resemblance to our own considerations of schedules of probability, but whereas the units of relevance for each schedule in our account are to be determined by definition of the actual state of the universe at that stage,[151] Fisher's units are possible states of nature defined in the m − space.[152]

McShane then also invites readers to see that we can reject the (still-) influential but mistaken abstraction wherein "one speaks more easily of gene frequencies and fluctuations."[153] We are invited, instead, to keep our sights on concrete circumstances that include, for instance, patterns of predation, temperature, and organic processes. What is needed is a

a principle which would bring together an anticipation of numbers and distributions and kinds and species and of reasons for the frequencies of kinds, one must include from the beginning a view of classical laws as they are realized in the concrete. **[And so, the] succession of schedules of probability must have for defined units not species, nor individuals, nor populations, nor possible world states, but types of concretely possible schemes.**[154]

McShane goes on and invites readers to share in formulating a concretely informed classical-statistical heuristics of world process called *emergent probability*, consisting of

sequentially-dependent recurrence-schemes. By taking these schemes, or the series of them—the earlier conditioning the later—in conjunction with their respective probabilities, we reach a view of emergence or evolution as the realization of a conditioned series of recurrence-schemes in accordance with a succession of probability schedules. Finally, as we already noted, the schemes may contain, or be contained within individual organisms. Therefore concomitant with the emergence of schemes there is the emergence of different types of things.[155]

Note also that this heuristics leads to an affirmative answer to the question of whether or not evolution has "direction":

There are clearly 'evolutionary traps and blind alleys' and so it is best to stress the word 'overall' in the answering of the question. And in

---

[151] See note 153.
[152] *RSE* (1970), 238; page 197 below.
[153] *RSE* (1970), 238; page 197 below.
[154] *RSE* (1970), 238; page 197 below.
[155] *RSE* (1970), 239; page 198 below.

this sense our answer is immediate: the direction of evolution is as a matter of fact a direction of increasing systematization. That increase of system is built into the principle of emergence enunciated above, since prior schemes condition later schemes.[156]

Here, then, McShane has achieved the main goal of the book which, it turns out, is a verifiable open explanatory heuristics of world process. "Evolution is not fully explained, but the principle of emergent probability [was, and remains] the best possible contemporary anticipation of that explanation."[157]

3.4 Growth vectors in *RSE*

In this subsection, I draw attention to some aspects of the book that were not highlighted as main goals but that either explicitly or implicitly call for development.

(a) There are "levels of inquiry"[158] and "levels of things."[159] Part of what is needed is an "apt symbolism."[160]

(b) "[T]he notion of central form is consistently passed over: its exclusion will be evident to those familiar with Lonergan's position; its inclusion would have been a stumbling block to those of other schools."[161]

(c) "As is evident from most of the examples, the cyclic or circular nature of the recurrence is not physical but metaphorical. Still, it is a metaphor in harmony with long usage."[162]

(d) "Since therefore, if any of the set [of a recurrence scheme] occurs, all occur, the change of circumstances would seem to have effected something like a jump in probability from the product to the sum of the particular probabilities."[163]

(e) The approach of F. M. Fisher "to produce 'a rather grandiose picture of history;'"[164] and "[t]he concrete process of evolution … involves a set of levels of non-systematic or random processes."[165]

(f) Growth, development, and progress: In in Halifax (circa 1986), McShane told me that his original idea for his PhD was that it

---

[156] *RSE* (1970), 241; page 200 below.
[157] *RSE* (1970), 245; page 203 below.
[158] *RSE* (1970), 170; page 141 below.
[159] *RSE* (1970), 8; page 6 below.
[160] RSE (1970), 89, 251; pages 73, 207 below.
[161] RSE (1970), viii; page lxiv below.
[162] *RSE* (1970), 227; page 188 below.
[163] *RSE* (1970), 230–231; page 191 below.
[164] *RSE* (1970), 237–238; page 196 below.
[165] *RSE* (1970), 246; page 204 below.

be a contribution to a scientifically grounded heuristics of development. He was dissuaded from the plan, on the basis that it would have been too large a project for the thesis. Evidence of that interest, however, is found throughout *RSE*. From chapter 1 onward, the problem is touched on explicitly, 126 times. Significant progress is made, and *development* is discussed with increasing frequency in the last two chapters so that, for instance, on pages 210–211 of *RSE* (1970), [166] the word 'development' is used 15 times. There are also eight loci of text in the book that regard "progress" and 25 regarding "growth." Development, growth, and progress are discussed in various contexts, including organic, psychological, intellectual, mathematical, scientific, human, social, societal and historical.

(g) With an evident control of meaning, McShane does not confuse the problem of understanding "economic system"[167] with that of understanding "rational behavior." [168] Acknowledging the existence of "economic cycles,"[169] we see evidence of his familiarity with views of the day, as well as his empirical orientation. That orientation is further revealed when, in the context of discussing Fisher's view on Markov processes, he acknowledges the potential "value in certain detailed investigations of economics."[170] However, no further hints are provided in *RSE*. In the spring of 1996, McShane wrote: "As far as I recall, it was in 1968 that I received two postcards, on two consecutive days, from Lonergan. The first merely asked if I knew of an economist that might be interested in tackling his 1944 economic work. The second gave the reason for the first: he had been reading Metz and felt it was high time that an economics emerged based on something more than disputes about the family wage. At all events, he sent me the 1944 economics, and I gave it sufficient attention to sense its remoteness."[171]

---

[166] See pages 174–175 below.

[167] *RSE* (1970), 210; page 174 below.

[168] *RSE* (1970), 58; page 46 below.

[169] *RSE* (1970), 226; page 188 below.

[170] *RSE* (1970), 237–238; page 197 below.

[171] "Work in Redress: The Value of Lonergan's Economics for Lonergan Students," *The Redress of Poise*, p. 2, http://www.philipmcshane.org/website-books, (accessed December 1, 2020). See also *Bernard Lonergan: His Life and Leading Ideas*, p. 103, and Frederick Lawrence (ed.), "Editors' Introduction," *Macroeconomic Dynamics: An Essay in Circulation Analysis*, University of Toronto Press, 1999, p. xl.

(h)   McShane refers to Toynbee's Theory of History, as well as Goudge's question about "concrete history."[172] On the possibility of synthesizing historical and systematic explanations, McShane notes that "[t]here are deeper difficulties involved here that we must pass over: cf. B. Lonergan's forthcoming *Method in Theology* on the question of history."[173]

4.   Beyond 2020: Results emergent or imminent in *RSE*

Section 3.4 lists a few results in *RSE* that invite development. For (a), (b), (f) and (g), one may look to the opera omnia of Philip McShane.[174] Following up on (g), for example, there is the need to understand, develop, and implement the economics discovered by Lonergan.[175] While not a stated purpose of *RSE*, the book includes significant advances in heuristics of growth and development that, by the same token, reveal directions for future progress. In this section, I focus on (c), (d), (e) and (h), under three main discussion points, [1], [2] and [3]. I selected these because they are nascent in *RSE*; regarding statistical method, they have commonalities; and, thus far, they have remained largely untouched in the

---

[172] *RSE* (1970), 237–238, 244; pages 196, 203 below.

[173] *RSE* (1970), 244, note 38; page 203 below.

[174] See note 21.

[175] Michael Shute, *Lonergan's Discovery of the Science of Economics*, University of Toronto Press, 2010. See *Macroeconomics Dynamics: An Essay in Circulation Analysis*, CWL 15 and *For a New Political Economy*, CWL 21. See also: Philip McShane, *Piketty's Plight and the Global Future*, Vancouver, Axial Publishing, 2014 and *Economics for Everyone*, 3rd edn., Vancouver, Axial Publishing, 2017; *Do You Want a Sane Global Economy?* (note 65 above); Bruce Anderson and Philip McShane, *Beyond Establishment Economics*, Nova Scotia, Axial Press, 2002; Bruce Anderson, "Basic Economic Variables," *Journal of Macrodynamic Analysis*, vol. 2, 2002, pp. 37–60; Terrance Quinn, "Lonergan's Contribution to Ecological Economics," *Ecology, Economy and Society* (forthcoming in 2021), https://ecoinsee.org/journal/ojs/index.php/ees/index; James Duffy, "Minding the Economy of *Campo Real*," *Divyadaan: Journal of Philosophy and Education*, vol. 29, no. 1, 2018, pp. 1–24; and Paul St. Amour, "Situating Lonergan's Economics in a Context of Collaboration," *Lonergan Workshop*, vol. 23, 2009, pp. 423–443.

literature.[176] My immediate purpose is modest, namely, to share preliminary searchings regarding "the view that would result from developing"[177] these ideas.

[1] "Under the unfavourable circumstances mentioned the members of the set may be said to be practically unrelated, and so the probability of occurrence of the whole set can best be represented by something like the product of the probabilities of the members. But if the circumstances change—in our simple parallel, something akin to a pressure-reduction in the gas—then the set of initial conditions are no longer unrelated. If the earth comes into any one of the set of conditions involving the suitable position and velocity for the particular elliptic orbit, all other members of the set defined by that elliptic orbit will occur. **Since therefore, if any of the set occurs, all occur, the change of circumstances would seem to have effected something like a jump in probability from the product to the sum of the particular probabilities.** Needless to say, our example is extremely simple and so not representative of concrete problems of calculating probabilities in evolution theory. But at least one can appreciate clearly in this instance what will be less clear and precise in more complex instances, the change in the probability of emergence on the fulfilment of prior conditions or the provision of adequate environment."[178]

[2] "The principle of evolution or of emergent probability towards which we are moving is ... not some abstract principle of selection for an aggregate of possible worlds. It is geared to the empirical investigation of the success-sive stages of the actual process of emergence and evolution."[179] Certain aspects of the approach suggested by F.M. Fisher were identified as being

---

[176] The question in (d), however, is raised in Duffy, "El azar, la probabilidad emergente y la cosmopolis." An attempt to summarize (d) is given in Frank E. Budenholzer, *Zygon*, vol. 39, no. 2, 2004, p. 346. But the discussion there is in general terms, attempts summary views, and is at a considerable remove from scientific practice. Regarding the dangers of "summary," see note 17. McShane spent much of the next 50 years developing heuristics that eventually sublated "work on Fisher and Markov into a flow of world maps that, at say, various intersections of latitudes and longitudes, has a statistics of recurrence-schemes of progress and probable 'situation room' components of progress." Philip McShane, *Interpretation from A to Z*, Vancouver, Axial Publishing, 2020, 147. This is subsumed by the symbolization $\{M(W_3)^{\Theta\Phi T}\}^4$. McShane, *Interpretation from A to Z*, pp. 146–150. Regarding the possibility of interpreting the meaning of McShane's works, see note 21.

[177] *Method in Theology* (1972), 250; CWL 14, 235.

[178] *RSE* (1970), 230–231; page 191 below. See also CWL 3, 144. McShane's description of the problem is looser. But an empirical open heuristic is implicit in Lonergan's prose.

[179] *RSE* (1970), 237; page 196 below.

inadequate.[180] *RSE* then moves on to a heuristics of emergent probability. **However, at the discussion of Fisher's work, we are at a high point of *RSE* in the sense that the discussion tacitly hints at the need to possibly retain aspects of Fisher's approach, for "some kind of multiple Markov matrix as giving the required picture of history."**[181] Part of the challenge is that we will need to incorporate schemes of recurrence discovered in science (including human science). The "introduction of the scheme as unit would complicate considerably the statistics. … Still, difficulty is not a criterion of scientific unacceptability."[182]

[3] McShane recalls Toynbee's *Theory of History*, as well as Goudge's question about "concrete history."[183] On the possibility of synthesizing historical and systematic explanations, McShane notes that "[t]here are deeper difficulties involved here that we must pass over: **cf. B. Lonergan's [then forthcoming]** *Method in Theology* **on the question of history."**[184]

For readers already competent in modern methods of applied statistical analysis, a heuristic something like [1] may have an immediate plausibility.[185] McShane was strategically non-specific in his statement. Again, it will be for empirical investigators to discover and work out appropriate measures and probability distributions for emergence of particular recurrence-schemes (classes of which also are to be determined). Still, more detail would be helpful. My purpose in section 4.1, then, is to shed some light on the plausibility of the heuristic in [1] as well as on the fact that it will be essential to progress in evolutionary studies.

Regarding [2], one of McShane's introductory examples is of a dog that chases a rabbit. The rabbit flees and hides in a burrow. When the dog leaves, the rabbit comes out. Later, another dog chases the rabbit. And so on. There is a scheme of recurrence.[186] But grass may also contribute to a rabbit's dietary cycles. And "components" of other "overlapping" schemes of recurrence are found in photosynthesis that occurs in grass in the field in which both dogs and rabbits run. Mediated by the earth's atmosphere, grass conjugates with, among other *things*,[187] the sun's vast numbers of physico-chemical compounds (the sun, it

---

[180] *RSE* (1970), 237–238; pages 196–197 below. See in this introduction pp. xxix, xxxi, xxxii, xliii, and xlvii.

[181] *RSE* (1970), 237; page 196 below.

[182] *RSE* (1970), 238–239; page 197 below.

[183] *RSE* (1970), 237–238, 244; pages 196, 203 below.

[184] *RSE* (1970), 244, note 38; page 203 below.

[185] See also CWL 3, 144.

[186] *RSE* (1970), 224; page 185 below. See CWL 3, 141–143.

[187] CWL 3, ch. 8. See also Meghan Allerton and Terrance Quinn, "The Notion of a Thing," *Journal of Macrodynamic Analysis*, vol. 14, 2020, pp. 95–109.

seems, has something of the order of 1.2×10^{57} molecules) and that way is involved in the world's oxygen and carbon cycles. And let us not forget the "non-systematic"[188] Great Asteroid Belt. And so on. That is, vast ranges of countless flexibly "linked" and "overlapping" schemes of recurrence are constitutive of more than 8.7 million species currently on earth and all things in the emergent universe of two trillion or more galaxies.[189] Indeed, we too are part of the process that "brings forth its own unity in the concentrated form of a single intelligent view."[190]

In section 4.2, I discuss the need and possibility of developing mathematical statistics for the emergence of schemes of recurrence. Two parts of the problem will require major advances: (1) As already partially established in *RSE*, there is a need to identify components of schemes of recurrence explanatorily and luminously. This will involve a heuristics of "aggreformic things." (2) In order to obtain statistical structures for schemes, we will need to work out something like "spaces of schemes of recurrence with appropriate topologies." These will need to be developed from and be verifiable in subsets of schemes and "sub-schemes," as determined by concrete circumstances. I do not attempt to suggest specific structures. Such progress will be the fruit of future science increasingly operating within generalized empirical method. That new stage in science and philosophy eventually will be "functional," which leads to [3].

In section 4.3, I follow up on [3] which points to Lonergan's breakthrough to "functional specialization." This is not "emergent" in *RSE*, as such. But it is hinted at in the footnote quoted, and it also is part of my present heuristics within which I am discerning implications of *RSE*. As already mentioned in the Preamble, building on results from sections 4.1 and 4.2, I draw attention to some elementary heuristics regarding timescales and structurings in probabilities of emergence of schemes of recurrence *in functional communications* $C_{ij}, i, j = 1, 2, \ldots, 8, 9$.[191]

---

[188] This is a fundamental notion throughout *RSE* (1970), as is "systematic process." See also CWL 3, 71–76, 79, 121, and Index, under 'process,' 'development' and 'emergence.'

[189] Christopher J. Conselice et al., "The Evolution of Galaxy Number Density at z < 8 and its Implications," *The Astrophysical Journal*, vol. 830, no. 2, 2016, pp. 83–100. http://dx.doi.org/10.3847/0004-637X/830/2/83, (accessed December 1, 2020). CWL 3, 544.

[191] Philip McShane, *Interpretation from A to Z*, p. 26 and *A Brief History of Tongue: From Big Bang to Coloured Wholes*, p. 108.

4.1 Probabilities of emergence of schemes of recurrence

Recurrence-schemes were introduced in section 3.2.[192] For present purposes, I adjust notation slightly from that given in *Insight*. Instead of events "$A, B, C, ...$",[193] I write $E_1, E_2, ..., E_n$, $n \gg 1$. For a *scheme of recurrence*, "a series of events $E_1, E_2, ..., E_n, n \gg 1$ would be so related that the fulfillment of the conditions for each would be the occurrence of the others. Schematically, then, the scheme might be represented by the series of conditionals: if E_1 occurs then E_2 will occur; if E_2 occurs then E_3 will occur; ...; if E_{n-1} occurs then E_n will occur; and if E_n occurs then E_1 will recur. Such a circular arrangement may involve any number of terms, the possibility of alternative routes, and in general any degree of complexity."[194] I denote such a recurrence-scheme by $(E_1 E_2 ... E_n)$.[195]

In [1], you may wonder why there is the qualifier "something like." Recall that the focus of *RSE* is on "pre-formulational"[196] aspects of the problem. The replacement of "something like" with a particular measure and probability distribution will be the work of empirical investigators. Note also that **the problem is about situations where circumstances not only have changed but may be in the process of ongoing change.** And so, measures and probability distributions needed also may be in transition. In order to verify *probabilities of emergence of a scheme of recurrence*, investigators will need to determine **relative actual frequencies of emergence of a scheme of recurrence**, work out and make use of empirically-appropriate measures and, if possible, identify and verify ideal relative frequencies of emergence, types of randomness, error bounds, and so on, as required. If circumstances are changing, results might need to be updated. This might include the inclusion of new events and new schemes of recurrence. I say all of this with the understanding that the possibility of such results is "remote"[197] to present-day inquiry. I will return to this point at the end of subsection 4.1. **When a scheme of recurrence becomes concretely possible,** the heuristic in [1] partly is to the effect that the empirical probability of emergence of schemes of recurrence of the form $S = (E_1 E_2 ... E_n)$ typically is greater than products of probabilities of *scheme-initiating events $E_1, E_2, ..., E_n$.*[198]

---

[192] See notes 130–132. They are defined in *RSE*, 219; page 182 below; and in CWL 3, 141–143.

[193] CWL, 3, 141.

[194] CWL, 3, 141.

[195] The notation is inspired by the notation for "cycles" in elementary group theory. But this is not to suggest a classical group structure.

[196] *RSE* (1970), 151, 155, 157; pages 125, 128, 130 below.

[197] *RSE* (1970), 230–231; page 191 below.

[198] Note that, for each $i = 1, 2, ..., n,$ the set of scheme-initiating events E_i is a subset of all events of the form E_i. It is assumed that occurrences of events E_i that are potentially scheme-initiating can be empirically distinguished from events E_i of the same form that are consequent to events E_j within an already-functioning scheme of

This will be of significance in evolutionary studies because once a scheme of recurrence is established, then, other things being equal (and statistically speaking), the ongoing recurrence of (aggregates of) events $E_1, E_2, ..., E_n$ in the scheme is assured. By the same token, once a scheme of recurrence is established, emergence is no longer an issue, let alone probabilities of emergence.[199]

Notice, then, that [1] tacitly refers to two (or more) sets of probabilities. The heuristic assumes that initially there are probabilities $prob^{initial}(E_1), prob^{initial}(E_2), ..., prob^{initial}(E_n)$, for events of the form $E_1, E_2, ..., E_n, n \gg 1$. But "circumstances change"[200] in such a way that *emergence of schemes of recurrence* of the form $S = (E_1 E_2 ... E_n)$ becomes concretely possible. **The statement regarding "the sum of the particular probabilities"**[201] **is in reference to the potentially *new* probabilities $prob(E_1), prob(E_2), ..., prob(E_n)$ of *scheme-initiating events* in the *new or changed* circumstances.** Or again, the heuristic regards how the quantities $prob(E_1), prob(E_2), ..., prob(E_n)$ (*of scheme-initiating events in the changed circumstances*) relate to *prob* (emergence of recurrence-schemes S) of the form $S = (E_1 E_2 ... E_n)$.

To help flesh this out somewhat, I start with a numerical example. In the example, the goal is to obtain an approximation for the empirical probability of massive objects (e.g., asteroids) falling into elliptical solar orbit along the orbital plane of the solar system. Immediately, however, you may wonder in what sense circumstances have changed. In this example, large timescales are needed. The sun was not always present. And for "empty space," other things being equal, based on current knowledge, the probability of aggregates of objects of small mass relative to the sun regularly intersecting locations in an elliptical toroid (a 3-dimensional elliptical tube) is a large product of small probabilities of mainly independent events. As mass distributions in space confirm, such a product is effectively zero. With a central mass like the sun present, however, a gravitational field makes the emergence of solar orbits concretely possible.

As I present it here, the numerical part of the problem is artificial, but it will help introduce some of the quantities involved. Suppose, then, that observational telescopes gather data daily, along eight non-overlapping solid angles along the solar orbital plane. Events $E_1, E_2, ..., E_8$ will be massive objects observed along each of the eight solid angles, respectively, together with, in each case, velocities and radial distances from the sun. The statistical inquiry is into the emergence of

---

recurrence. Otherwise, the mathematical formulation breaks down, and verification the heuristic is not possible. The meaning of "within" is determined by the scheme recurrence.

[199] Inquiring into frequencies, densities, and such of already established schemes of recurrence are further questions. See notes 141–145.

[200] See note 178.

[201] See note 178.

schemes of recurrence of the form $S = (E_1 E_2 ... E_8)$. In this example, the key to the problem is that there are eight possible types of event that can initiate the same type of scheme.

Suppose that observational intervals are "per day" and that, from a sample of, say, $N = 300$ days, on each day either a scheme of recurrence emerged, or not. (To determine whether a scheme of recurrence emerged would, of course, require further data as well as computations using equations of motion (e.g., Newtonian, or Einsteinian).) In the example, suppose that the total number emergences of schemes of recurrence of the form $S = (E_1 E_2 ... E_8)$ is $m_S = 200$. Source data might be something like the following:

	Day 1	Day 2	Day 3	Day 4	Day 5		Day 300
Led to solar orbit	N/A	Yes	Yes	No	Yes	...	
Observed mass and velocity	0	1	1	1	1	...	
Axis	N/A	3	5	7	3	...	
Event type	N/A	E_3	E_5	N/A	E_3	...	

To tally experimental results, let m_i be the number of times a mass-velocity of type E_i falls into solar orbit in the orbital plane (in other words, initiates a scheme of recurrence $S = (E_1 E_2 ... E_8)$.).

For example, results might be $m_1 = 20$, $m_2 = 35$, $m_3 = 17$, $m_4 = 54$, $m_5 = 7$, $m_6 = 14$, $m_7 = 23$, $m_8 = 30$; and in the trial of $N = 300$ days, $m_S = 200 = m_1 + m_2 + \cdots + m_8$. The relative actual frequencies would, therefore, satisfy

$$\frac{200}{300} = \frac{m_S}{N}$$
$$= \frac{m_1 + m_2 + \cdots + m_8}{N}$$
$$= \frac{m_1}{N} + \frac{m_2}{N} + \cdots + \frac{m_8}{N}$$
$$= \frac{20}{300} + \frac{35}{300} + \frac{17}{300} + \frac{54}{300} + \frac{7}{300} + \frac{14}{300} + \frac{23}{300} + \frac{30}{300}$$

Continuing in this way, and supposing that with enough samples investigators succeed in determining probabilities, it is evident that the probability of emergence of schemes of recurrence of the form $(E_1 E_2 ... E_8)$ would be something like a sum of probabilities, that is,

$$prob\big(\text{emergence of recurrence-schemes } (E_1 E_2 ... E_8)\big)$$
$$\approx prob(E_1) + \cdots + prob(E_8). \tag{1}$$

If you are familiar with the elementary mathematics here, you will see that essential features of the example generalize. Suppose that observational methods allow for the detection of zero or one event (or unit aggregate of events) E_i not already in a scheme of recurrence per observational interval or occasion; and that it is possible for investigators to determine which such events (or aggregates of events) initiate a scheme of recurrence $(E_1 E_2 \ldots E_n)$.[202] Suppose that in a sample of N observational intervals or occasions, m_S is the number of schemes of recurrence that emerge and that m_i is the number of (recurrence) scheme-initiating events (or aggregates of events) of type E_i. As in the example, we get that $m_S = m_1 + m_2 + \cdots + m_n$, from which it follows that the relative actual frequency of emergences of schemes of recurrence is a sum of relative actual frequencies of events from which schemes of recurrence emerged. That is,

$$\frac{m_S}{N} = \frac{m_1}{N} + \frac{m_2}{N} + \cdots + \frac{m_n}{N} \tag{2}$$

Again, assuming that investigators are successful in determining relevant empirical probabilities, to some approximation, we can expect that

$$(prob\ emergence)(E_1 E_2 \ldots E_n)$$
$$= prob(E_1) + \cdots + prob(E_n). \tag{3}$$

The heuristics of [1] refer to "something like a jump in probability from the product to the sum of the particular probabilities." **Note, however, that the jump refers to mathematical form and need not regard the same probabilities. For not only have circumstances changed but the probabilities in question now regard emergence of schemes rather than events E_i.** The way in which these two sets of probabilities relate will be an empirical problem determined by situations. We can, however, relate the product of probabilities $prob(E_i)$ to their sum, by fairly obvious inequalities. To see that, observe that since empirical probabilities are bounded below by zero and above by unity, for each

$$i = 1, 2, 3, \ldots, n, \quad prob(E_1) \cdots prob(E_n) \ll prob(E_i).$$

And so,

$$prob(E_1) \cdots prob(E_n) \ll \sum_{i=1}^{n} prob(E_i).$$

The heuristic of [1] regards events (or aggregates of events) that initiate schemes of recurrence. In practice, more than one scheme of recurrence might be observed to emerge per observational interval or occasion. In such cases, investigators may need to look to the emergence of k schemes of recurrence, $k = 0, 1, 2, 3, \ldots$, in each of N observational intervals. (The statistician will see the histogram lurking.) The next few paragraphs touch on this problem. If you are not familiar with the mathematics needed, you might go directly to the paragraph (below) that begins "Now, these heuristic equations are well and good. However …"

---

[202] Regarding notation, see note 195.

It is because of the recurrence-scheme structure that the emergence of k schemes of recurrence in an interval or occasion can be initiated in many ways. To make this more precise, let $m_1 = e_1$ mean that e_1 events of type E_1 *initiated* schemes of recurrence; and so on, for $m_2 = e_2,\ldots$, and $m_n = e_n$. (For example, $k = 2$ schemes of recurrence emerge in an interval or occasion if any of the following combinations occur: $(m_1 = 2, \; m_2 = 0, \ldots, m_n = 0)$, or $(m_1 = 1, \; m_2 = 1, \ldots, m_n = 0)$ or $(n_1 = 0, \; m_2 = 2, \; m_3 = 0, \ldots, m_n = 0)$, and so on.) It follows that k schemes of recurrence emerge in an interval or occasion if and only if $e_1 + e_2 + \cdots + e_n = k$ in that interval or occasion. Mathematically, the number of ways of obtaining k schemes of recurrence is the number of partitions of k of length n. This is a much-studied function in modern number theory. But which cases occur and initiate a scheme of recurrence of the required form is determined through empirical inquiry. To some approximation, then,

$$prob(\text{emergence of scheme } \mathbf{S}, k\text{–times})$$
$$\approx \sum_{e_1+e_2+\cdots+e_n=k} prob(m_1 = e_1 \text{ and } m_2 = e_2 \text{ and } \ldots \text{ and } m_n = e_n), \qquad (4)$$

where the sum is across all combinations $e_1 + e_2 + \cdots + e_n = k$ that occur, and summands correspond to scheme-initiating events. Consequently, the empirical probability of emergence of (exactly) k schemes of recurrence $S = (E_1 E_2 \ldots E_n)$ per observational interval or occasion again will be something like a sum of joint probabilities.

Where equation (4) follows from the heuristic in [1], it also leads to a parallel result. That is, **in the new situation**[203] (in which schemes of recurrence are emerging), the probability of emergence of k schemes of recurrence (in intervals or occasions) also will be "something like a jump"[204] but this time to a sum of products.

To see this, recall that one of the hypotheses throughout these derivations is that **occurrences of events E_i not already in a scheme of recurrence** are "practically unrelated."[205] And so, at least for a time, summands in equation (4) will be something like products. That is, equation (4) will be something like

$$prob(S \text{ emerges } k \text{ times})$$
$$\approx \sum_{e_1+e_2+\cdots+e_n=k} \big(prob(E_1)\big)^{e_1} \big(prob(E_2)\big)^{e_2} \cdots \big(prob(E_n)\big)^{e_n}. \qquad (5)$$

---

[203] "[C]ircumstances change" (note 178).

[204] See note 139.

[205] See note 139. Typically, this means that, to some approximation, they are statistically independent. This is often the case when, for example, we are considering mass-velocities at relatively remote locations in the heliosphere or, say, biochemical events at relatively remote locations on earth.

Now, these heuristic equations are well and good. But they provide merely a glimpse of richness and complexities to be expected in even the most elementary cases. It is also true they barely touch on what will be needed in statistical analysis of emergence of schemes of recurrence.

For one thing, the equations are derived under a special hypothesis, namely, that for a scheme of recurrence that has emerged, (relative) frequencies of events (or aggregates of events) E_i that *initiate* schemes of recurrence can be determined. But to determine (or estimate) such frequencies will certainly be challenging in, for instance, the study of the gradual emergence of the TCA cycle in organisms in global ecosystems. However, empirical challenges aside, the hypothesis is legitimate because for the emergence of every scheme of recurrence $(E_1 E_2 \dots E_n)$ there are scheme-initiating events (or aggregates of events) E_i.

Note also that, in the illustrations provided, computations have been in the discrete case, with the "counting measure." In modern science, different measures are needed (in particle physics, astrophysics, thermodynamics, biochemistry, soils science, Brownian motion, Wiener processes and other diffusion processes, multispecies population dynamics, sociology, and so on). Mathematical developments notwithstanding, empirical probability functions, states, as well as types of randomness are determined by "genera and species" of events or occurrences in concrete situations. Future empirical inquiry into the emergence and functioning of schemes of recurrence will need a heuristics that includes vast ranges of types of probability functions defined in terms of integrals ("sums") and measures that, as determined by investigators, are probably appropriate—be they discrete, continuous, or otherwise. Undoubtedly, these also will vary along schemes of recurrence, according to the various types of event E_i.

Furthermore, the heuristic [1] is open. It regards statistical emergence of schemes of recurrence that may be systematic processes or non-systematic processes.[206] In the elementary example, I discussed elliptical solar orbits. Each such orbit approximates being a systematic process. But it is non-systematic processes that are most prevalent in world process. In such cases, the meaning of the series of conditionals "if E_i occurs then E_{i+1} will occur" is obviously statistical, and "circularity" is just a metaphor.[207] Also, the "scheme-initiating" E_i being "*practically* unrelated"[208] begs the question. As is now known in, for example, biochemical pathways in organisms, multi-species cycles in ecosystems, and global chemical cycles, typically there are "cross-over influences." And so, in many cases, statistical analysis of emergence of schemes of recurrence also will need to include statistical data for determining conditional probabilities

---

[206] See note 108.

[207] See note 133.

[208] See note 178.

$prob(E_i|E_j), i, j = 1, 2, ..., n$. In other words, sequences of some kind of stochastic or Markov-like matrices will be needed.

The first part of this section provides a few details regarding the heuristic that, once concretely possible, probabilities of emergence of a recurrence-scheme $(E_1 E_2 ... E_n)$ will be something like sums across probabilities of recurrence-scheme-initiating events E_i. While there is already compelling evidence from ongoing scientific progress, empirical inquiry into probabilities of emergence of recurrence-schemes, as such, remains a future possibility. As McShane pointed out in *RSE*, it is still true that "[t]he notion is very general and heuristic and the possibility of filling out its details [remains] remote."[209] But the work of this section has been fruitful. In addition to fleshing out an elementary case somewhat, what is now highlighted is the need for complex, nuanced heuristics for statistical analysis of schemes of recurrence in world process.

You may notice that I did not mention influences from "overlapping" schemes of recurrence. That brings me to [2] which, among other things, regards sets of schemes of recurrence.

4.2 Emergent probability: the total quasi-Organism[210]

As a heuristics, emergent probability generates "hypotheses, not deductively, but by filling out the open structure through empirical investigation. ... [The principle] represents the anticipation of the best that can be achieved in accounting for world process. ... The concrete process ... involves a set of levels of non-systematic or random processes ... [And] the best that one can do toward an integrated view is to establish a schedule of probabilities for the immediate course of events at that stage."[211] Emergent probability is, then, a middle term for an Aristotelian syllogism.[212] In view of several decades of scientific progress, might we add to the heuristic? As in seasonal "salmon runs," Fisher's leap[213] fell short. But he was on to something.

Clues are available in more than a century of developments in applications of Markov theory and applied mathematical biology. But again, results will need to be orientated "from within."[214] By contrast, systems theories and "complexity" do not solve the problem of "levels" (identified in *RSE*); nor do they adequately account for growth, development, and emergence. As in Rashevsky's work,[215] positive results can be obtained by selecting subsets of variables and working with

---

[209] *RSE* (1970), 231; page 191 below.

[210] Pierrot Lambert and Philip McShane, *Bernard Lonergan: His Life and Leading Ideas*, p. 250.

[211] *RSE* (1970), 246–247; page 204 below.

[212] *RSE* (1970), 247; page 204 below.

[213] See note 152.

[214] *RSE* (1970), vii; page lxiii below.

[215] See section 3.1.

averages and probabilities. And Avner Friedman's observation from 2010 still applies: "Even the most successful models … deal only with limited situations, ignoring all but the most essential variables."[216] With fewer variables in play, using geometry, topology, differential equations, combinatorics, analysis, methods for stochastic processes, and so on, contemporary models are developed to investigate aggregates and averages.[217] However, averages and probabilities do not explain the concrete—such as an amoeba reaching for a food particle, or a salmon leaping upstream. Also note that, in contemporary models, components of the models are in a sense too "random"; or rather, semi-random. For, I do not mean that Markov matrices have not been enormously helpful. So far, however, events, occurrences, and (statistical) states mainly are defined non-luminously.

Within this historical context, my purpose in this section 4.2 is to point to progress needed in order to take advantage of the identification obtained in *RSE* that "units of evolution" are concretely operable classical schemes of recurrence.[218] Foundational results reveal that there is a need for new mathematical and statistical methods for sets of overlapping schemes of recurrence, sets of overlapping sub-schemes of recurrence, and components of schemes and sub-schemes of recurrence.

To be sure, developing mathematics of schemes of recurrence will need to include progress in identifying components of schemes of recurrence. *RSE* helps bring out that the appropriate method will be generalized empirical method; and that part of what is needed is progress in (luminous) heuristics of "aggreformism" and "levels of things, the justification for the existence of different irreducible

---

[216] Avner Friedman, "What is Mathematical Biology and How Useful Is It?" *Notices of the American Mathematical Society*, vol. 57, no. 7, 2010, p. 852, https://www.ams.org/notices/201007/rtx100700851p.pdf, (accessed December 1, 2020). Of course, that begs the question. How does one know what "the most essential variables" are without an explanatory heuristics of the whole organism? Asserting that prescribed physical variables are "most essential" reveals an implicit operative reductionism. Again, though, the method yields positive results. For example, see Yizeng Li, Lingxing Yao, Yoichiro Mori, and Sean X. Sun, "On the energy efficiency of cell migration in diverse physical environments," *Proceedings of the National Academy of Sciences*, vol. 116, no. 48, 2019, pp. 23894–23900, https://doi.org/10.1073/pnas.1907625116 (accessed December 1, 2020), which focuses on two physical variables, "effective strength of focal adhesion and the coefficient of external hydraulic resistance" (23898). The results also reveal the need for more adequate heuristics and control of meaning.

[217] See also note 122.

[218] *RSE* (1970), 9–10; page 7 below. See also CWL 3, (9) on page 148 and (1) on page 149.

sciences."[219] A convenient heuristic symbolism is $f(p_i; b_j; c_k; z_l; u_m; q_n)$.[220] This will not be a matter of interfering with already effective symbolism. It partly will be a matter of reaching a new control of meaning while allowing for, as needed, *adaptations and developments* in expression. And so, as brought out in *RSE*,[221] for explaining organisms **"[t]here have to be invented appropriate symbolic images of the relevant chemical and physical processes; in these images there have to be grasped by insight the laws of the higher system that account for its regularities beyond the range of physical and chemical explanation; from these laws there has to be constructed the flexible circle of schemes of recurrence in which the organism functions; finally, this flexible circle of schemes must be coincident with the related set of capacities-for-performance that previously was grasped in sensibly presented organs."[222]**

Eventually, symbolism will be developed that promotes *linguistic feedback*: "At a higher level of linguistic development, the possibility of insight is achieved by linguistic feedback, by expressing the subjective experience in words and as subjective."[223] "[L]inguistic feedback is achieved … in the measure that explanations and statements provide the sensible presentations for the insights that effect further developments of thought and language."[224]

In physics, for example, we can imagine a "future Feynman" and collaborators, luminous in their talk of photons, electrons, and energy (dark or

---

[219] *RSE* (1970), 9; page 6 below.

[220] This symbolism for the results obtained in *RSE* (see, e.g., *RSE* (1970), 251; page 207 below) emerged later. See, for example, Philip McShane, *A Brief History of Tongue: From Big Bang to Coloured Wholes*, p. 119. An early version of the symbolism is in *Wealth of Self and Wealth of Nations, Self-Axis of the Great Ascent*, Hicksville NY, Exposition Press, 1975, p. 106. This book is now available at http://www.philipmcshane.org/published-books, (accessed December 1, 2020).

[221] *RSE* (1970), 250–251; page 207 below.

[222] CWL 3, 489. Some readers may recall the work of "Paul Humphreys, on 'emergence as fusion' in which events are unified multi-level events. His symbolism for hierarchies of levels $L_i < L_j$ (Timothy O'Connor and Hong Yu Wong, "Emergent Properties," in Edward N. Zalta (ed.), *The Stanford Encyclopedia of Philosophy*, 2020, "Emergence as 'fusion,'" 3.2.2, https://plato.stanford.edu/archives/sum2020/entries/properties-emergent) somewhat resembles symbolism provided in the present book. However, Humphrey's arguments are given in a context of philosophical discussion that becomes entangled in logical difficulties. … The logical structures $L_i < L_j$ are remote to individual living organisms, and do not draw on or connect with detailed empirical results in scientific practice." Terrance Quinn, *Invitation to Generalized Empirical Method in Philosophy and Science*, Hoboken, NJ, World Scientific, 2017, p. 149, n. 39.

[223] CWL 14, 85–86, note 55.

[224] CWL 14, 89.

otherwise[225]). Investigators will be making progress in identifying and distinguishing conjugate and central potencies, forms, and acts,[226] primary relations and secondary determinations, in contexts of up-to-date "gauge theory" (or whatever the standard model happens to be at that time). And it will be within a heuristics of "aggreformism," "genetic sequences," "flexible series," "successive stages," "integrator" and "operator" that biology will be growing in its understanding of, say, a dog-thing,[227] its anatomy, biochemistry, growth, development,[228] life cycle, and indeed, its "co-evolutionary" emergence with, among other things, human things.

Currently, there is a popular notion of "inter-face" of mathematics and biology. As noted above, however, much of what is being produced in contemporary mathematical biology is not verifiable in individual cells and organisms in vivo, in vitro or in situ. But that is not the whole story. [Successful] "work in mathematical biology is typically a collaboration between a mathematician and a biologist. The latter will pose the biological questions or describe a set of experiments, while the former will develop a model and simulate it." [229] Collaboration between mathematics and biology is now normal and, indeed, has historical origins. There are, for instance, methods, results, and goals of Sir James Lighthill (1924–1998). He was of the mind that "external and internal biofluiddynamics demand … interdisciplinary investigation. This requires always that a gifted biologist with a genuine interest in fluid-dynamics work in close collaboration with a gifted fluid dynamicist committed to problems arising in biology." [230] Lighthill's results may be dated. But I am drawing attention to Friedman's and Lighthill's operative heuristics. Both quotations hint of a fundamental need for a kind of mathematical biology whereby, instead of "interface," mathematics is *intrinsic* to biology. I am not suggesting that mathematics interfere with biology. Generalized empirical method is concrete and will attempt to advert to all relevant experience in biology.[231] More broadly, within a heuristics of aggreformism, there is "an invitation to mathematicians to

---

[225] P. Lambert and P. McShane, *Bernard Lonergan: His Life and Leading Ideas*, pp. 178–188. This will include verifiable heuristics of "layerings" of negentropies and entropies with (loosely speaking) "Boltzmann-type" results for "layerings" of forms.

[226] CWL 3, chs. 15–16.

[227] CWL 3, ch. 8. See also Meghan Allerton and Terrance Quinn, "The Notion of a Thing."

[228] CWL 3, 489.

[229] Friedman, "What is Mathematical Biology and How Useful Is It?" p. 852.

[230] James Lighthill, "Biofluiddynamics: A Survey," in A. Y. Cheer and C. P. van Dam (eds.), *Fluid Dynamics in Biology, Contemporary Mathematics*, vol. 141, 1993, p. 3.

[231] See notes 45–47.

explore the possibility of setting up series of deductive expansions that would do as much for [biology and] other empirical sciences as it has done for physics."[232]

Part of what is called for, then, is fundamental progress in luminously identifying components of schemes of recurrence. But there is a further challenge in the study of schemes of recurrence. Implicitly, the problem is not altogether unknown to modern scientific practice and results. If the name "scheme of recurrence" is not yet familiar (and, of course, the name as such is not important), a science of schemes of recurrence will begin with already known cycles and periodic processes, statistical or otherwise. Progress in luminous identification of components will reveal that known (and to-be-known) schemes of recurrence ascend and descend across and through vast ranges of "flexible levels or layerings" of things and their environments in world process. Some schemes of recurrence (e.g., some of the chemical schemes) are functional at multiple levels. Informed by empirical investigations, new mathematics will be needed for holding together sets of schemes, sub-schemes, and components of "micro, meso and macro" layerings of overlapping schemes of recurrence, in their primary relations and "secondarily-structured" secondary determinations. We will need empirically grounded topologies of sets of schemes determined within scientific practice, as well as ranges of measures and probabilities for sets of schemes (for their emergence, survival, and decay) that will be verifiable in vivo, in vitro, in situ and generally in concrete situations. In a "comeabout position,"[233] a scheme of recurrence is not a unity as such (as in the sense of potency, form, and act). A scheme does, however, have a unity evidenced in secondary determinations. And so, with topologies and layerings to be determined, a scheme is but a (scheme-) *section* of the total spatial temporal manifold that is world process, the total *quasi-Organism*.[234]

---

[232] CWL 3, 339.
[233] CWL 3, 537.
[234] See note 210.

Symbolism is needed.[235] Hypotheses of systems-theories notwithstanding, our experience is limited to spatial and temporal intervals (X, T). Let

$$\mathbf{M}_{PS(X,T)}\left(p_i;\ b_j;\ c_k;\ z_l;\ u_m;\ q_n\right)\left(\mathbf{H}(X,T)\right)$$

represent a range of schemes of recurrence S of the total quasi-Organism, known by persons $\mathbf{H}$ through spatial-temporal intervals (X, T), with P for verified probabilities of events, things, and schemes. There will need to be a control of meaning that, among other things, distinguishes aggreformic "levels" and "layerings" of central and conjugate potencies, forms, and acts. And, while emergent probability has direction,[236] there is no knowable limit to what the total quasi-Organism may yet be.

This is all so terribly brief. But it seems to me that the need for following up within some such heuristics emerges from and will be coherent with results in *RSE*. It will be a way to meet Fisher's hopes[237] and will accommodate McShane's rejection of specifics of Fisher's initial effort.[238] It invites detailed investigations in concrete situations. The symbolism also seems to fit with McShane's later metagrams.[239] In the present context, however, I am drawing special attention to schemes of recurrence; persons; and frames of reference. And, in basic position,[240] it will be increasingly (self-) evident that we and all are a unity in the emergent body of history called emergent probability.

---

[235] RSE (1970), 89, 251; pages 73, 207 below. CWL 7, 151 (150 in Latin).

[236] See note 156.

[237] See note 181.

[238] *RSE* (1970), 237–238; page 196 below.

[239] Philip McShane, *Brief History of Tongue*, p. 122; *Bernard Lonergan: His Life and Leading Ideas*, p. 161; and "Prehumous 2: Metagrams," p. 4, http://www.philipmcshane.org/prehumous, (accessed December 1, 2020). See also note 220 above.

[240] CWL 3, 413.

4.3 Progress-oriented schemes of recurrence

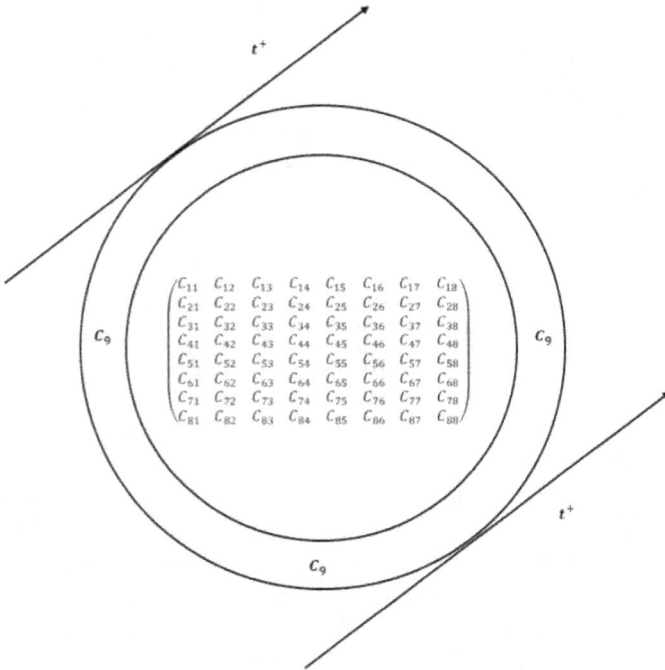

$$
\begin{pmatrix}
c_{11} & c_{12} & c_{13} & c_{14} & c_{15} & c_{16} & c_{17} & c_{18} \\
c_{21} & c_{22} & c_{23} & c_{24} & c_{25} & c_{26} & c_{27} & c_{28} \\
c_{31} & c_{32} & c_{33} & c_{34} & c_{35} & c_{36} & c_{37} & c_{38} \\
c_{41} & c_{42} & c_{43} & c_{44} & c_{45} & c_{46} & c_{47} & c_{48} \\
c_{51} & c_{52} & c_{53} & c_{54} & c_{55} & c_{56} & c_{57} & c_{58} \\
c_{61} & c_{62} & c_{63} & c_{64} & c_{65} & c_{66} & c_{67} & c_{68} \\
c_{71} & c_{72} & c_{73} & c_{74} & c_{75} & c_{76} & c_{77} & c_{78} \\
c_{81} & c_{82} & c_{83} & c_{84} & c_{85} & c_{86} & c_{87} & c_{88}
\end{pmatrix}
$$

Figure 1 "The matrix of academic collaboration is at a horizon-remove from streetmeaning."[241]

If in some way we share in Lonergan's Dream,[242] then we envisage a time when a dominant ethos of the academy will be progress for humanity, globally. Specializations called "disciplines" will be reorientated within *functional specialization*. Education will be increasingly informed by the Childout Principle, that is, "When teaching children geometry [or anything], one is teaching children children."[243] "[A]dverting to our conscious operations and to the dynamic structure that relates them to one another"[244] will be normal in all zones of inquiry. It will be a time when there is "resolute and [statistically] effective intervention in

---

[241] McShane, *A Brief History of Tongue*, p. 108.

[242] See note 33.

[243] Philip McShane, "Bridgepoise 8: New Beginnings in the Global Reaching of Lonergan," p. 6, http://www.philipmcshane.org/bridgepoise, (accessed December 1, 2020).

[244] CWL 14, 257.

the Dialectic" that is history.[245] Contexts of concern will be partly guided by a pointed question: What might the street-value be?

One of McShane's symbolizations for "functional communications"[246] is reproduced in Figure 1. As described by McShane, effectiveness of collaboration will be that of an "Archimedean screw,"[247] lifting cultures locally and globally.

Following on sections 4.1 and 4.2, questions arise. For instance, 'What is the probability of emergence of schemes of recurrence of functional communications C_{ij}?' And, 'How long will it take for such schemes of recurrence to emerge in history?'

As *RSE* brings out, the word *probability* is ambivalent. One type of answer can be through "v-probability."[248] In my assessment, eight functionally distinct tasks are (pre-) emergent in history.[249] A further judgement is that, Yes, I expect the eventual emergence of the new ethos, a flowering of humanity, a time when, while never without its problems, humanity "comes of age." In a "fanciful"[250] but reasoned account, McShane wrote about "Arriving in Cosmopolis" by about the year 9011 C. E.[251] At first blush, that may seem like a long time from 2021. Crises grip the world and human beings learn slowly. But humanity also hungers for saner times, and the earth's history sheds some light on timescales. It took about 80 million years for flowering plants to emerge; to eventually become widespread; and ultimately to dominate over conifers. That was in conjunction with ongoing "co-evolution" of countless multitudes of other things and organisms (e.g., bees) in vast ranges of emerging schemes of recurrence that, together, cumulatively-and-increasingly transformed global ecosystems. On geological timescales, the transition to "the age of flowering plants" was dramatic, less than 2% of geological time to-date.[252] What, then, of 9011 C.E., as suggested by McShane? And so, with McShane, might we not hope to do better than 80 million years? Might we not hope for a flowering of humanity that is widespread within 80 thousand months or so?[253]

---

[245] CWL 18, 305–308.

[246] McShane, *Interpretation from A to Z*, p. 26 and *A Brief History of Tongue*, p. 108.

[247] McShane, *Interpretation from A to Z*, p. 19.

[248] *RSE* (1970), 137–143; pages 114–118 below. See also CWL 3, 324–329, 574–575.

[249] See note 34.

[250] Philip McShane, "Arriving in Cosmopolis," p. 1, http://www.philipmcshane.org/website-articles, (accessed December 1, 2020).

[251] McShane, "Arriving in Cosmopolis," p. 1. See also *Method in Theology 101 AD 9011: The Road to Religious Reality*, Vancouver, Axial Publishing, 2012.

[252] The earth's age is approximately 4.54×10^9 years. The computation is $\left(\frac{80 \times 10^6}{4.54 \times 10^9} \right)$.

[253] With 80,000 weeks, can we beat McShane's estimate by about three centuries?

There is also "f-probability."[254] As in McShane's article,[255] I too will be "fanciful."[256] I envisage a reachable future when, in the academy, there is an emerging control of meaning; when there are increasing frequencies of scholarship and communications that approximate forms C_{ij}; when sequencings of mutually dependent C_{ij} increasingly are directed toward promoting progress-flows along the super-diagonal $C_{91}, C_{12}, C_{23}, ..., C_{78}, C_{89}$. (It is in C_{89} that the collaboration "bears fruit"[257] and without which the prior tasks "are in vain, for they fail to mature"[258].) It will be a time when, allowing for lags (months, years, or longer), there will be increasingly available statistics on sequences of communications "$C_{ij} \Rightarrow C_{kl}$". (The notation includes cases $i = j = k$.) In other words, I am envisaging a time when functional schemes of recurrence are becoming concretely possible. With global communications at least as good as they are now, we can anticipate a rising wave of interest in making "Archimedean-cyclic collaboration" a reality and, in particular, a growing commitment to getting results "to the streets."[259]

Initial emergences may be prolonged and will be enormously complex. But emergence also will be helped by that complexity. There will be diverse cultures, circumstances, needs and interests. Contributing to an emergent ethos of functional collaboration, then, there will be emerging ranges of differentiations of consciousness and horizon. There will be proclivities for distinct functional tasks, and all of this in slopes.[260] In brief, functional cyclic collaboration will be taking hold.

By that time, then, we can expect that empirical probabilities of emergence of functional schemes of recurrence will be something like sums (with ongoing changes in probabilities and conditional probabilities). Again, why "sums"? Allowing for ranges of interest and horizon (as just described), sources of functional schemes of recurrence increasingly will be from eight main groupings of tasks[261] and, generically, communication forms C_{ij}. The heuristics here is open, anticipatory of breakthroughs and genetic sequences, random shifts, increasing control of meaning, and so on. Of course, relative actual frequencies are not yet

---

[254] *RSE* (1970), 137–143; pages 114–118 below. See also CWL 3, 81–85.

[255] See note 251.

[256] See note 250.

[257] Lonergan, *Method*, 355; CWL 14, 327.

[258] Lonergan, *Method*, 355; CWL 14, 327.

[259] See note 247.

[260] Philip McShane, "Cantower 8: Slopes: An Encounter," http://www.philipmcshane.org/cantowers, (accessed December 1, 2020).

[261] See Figure 1, page xlviii above.

available. And so empirical probabilities of emergence of functional collaboration will be determined by scholars in that somewhat distant future.[262]

TERRANCE QUINN

TORONTO, CANADA
August 2021

---

[262] What can we do to promote the emergence of "scheme-initiating" events that, for a time, will be rare and random variations? How can we make progress in "kick-starting" the functional ethos? What is needed are humble beginnings in concretely focused self-revealing inquiry. There is a fundamental and pressing need for the *Duffy Exercises*. In these exercises scholars implement a three-step procedure that is pivotal to the fourth functional specialty that Lonergan calls *Dialectic*. See CWL 14, 235, lines 12–27; *Method in Theology* (1972), 250, lines 18–33. See also James Duffy, "Editor's Introduction," *Journal of Macrodynamic Analysis*, vol. 14, 2020, pp. 1–14. For further context, help, and inspiration, see also Philip McShane, chapter 3, "Self-Assembly," in *The Future: Core Precepts in Supramolecular Method and Nanochemistry*, Vancouver, Axial Publishing, 2019, pp. 41–57.

"The Riverrun to God: Randomness, Statistics, and Emergence"

Those interested in the original text of *Randomness, Statistics, and Emergence*, will find it here unchanged except for typo corrections and minor grammatical and bibliographic changes. What has changed is the context given by this Preface. But does not this change radically—and there's the radical rub—the meaning of the text of forty-five years ago? A question there for myself and my reader of any age, around which this Preface circles.

I pause over the addition to the original title, and draw your attention to the two first words, but now reversed, separated and decapitalized: "riverrun ... the." The two words, the first and last of James Joyce's *Finnegans Wake*, leave the four dots meaning that large strange flow of words that those two words bracket. And here I add to that large strange flow, and its global referents, some oddments of my own riverrun of the forty-five years since I completed that book.

The proximate field[1] of the riverrun is that contained in the recent book, *The Road to Religious Reality*, the title of which echoes a 21st century book on physics by Penrose, *The Road to Reality*.[2] *The Road to Religious Reality* has many messages, but a central message is that we, who make a study of religions, have to take seriously the powerful shift of the simplest of sciences, physics, that has occurred in the past two centuries.[3] By "taking seriously" I mean moving into that world in a way that carries one quite beyond the semipopular writing of Penrose, or, OM preserve us, *Scientific American*. That was my message to philosophy in *Randomness, Statistics, and Emergence*, and of course it was implicitly a message to theology and to my comrades in the Lonergan community. A second central message of *The Road to Religious Reality* is that we who study religions need to open

---

*Written in October, 2012. Updated references to Lonergan's Collected Works have been provided by the editor of this second edition of *Randomness*.

[1] My emphasis throughout here is on the concreteness of our minding reach, which Lonergan intimates by his use of the word field. "The field is *the* universe, but my horizon defines *my* universe." Bernard Lonergan, *Phenomenology and Logic*, vol. 18, Collected Works of Bernard Lonergan, Philip McShane (ed.), Toronto, University of Toronto Press, 2001, p. 199. My invitation is to a concrete fantasy that would lift scholarship into a full open content-full explanatory heuristic. The notes to the text weave round that neurochemical invitation.

[2] Roger Penrose, *The Road to Reality: A Complete Guide to the Laws of the Universe*, New York, Vintage Books, 2007.

[3] There is the deeper shift that places the question, What is physics? in the full cyclic dynamic from research to teachings and technologies. That topic cannot be developed here.

ourselves to the global lift in empirical studies that is given to us by Bernard Lonergan in his specification of a global dynamic collaboration.[4]

It is immediately important to locate the book *Randomness, Statistics, and Emergence* in the story of Lonergan's discovery of that wild dream. In the months before I began *Randomness*—titled "The Logic of Statistical Inquiry with Special Reference to Evolution Theory" in the Autumn of 1965 as a doctorate thesis in Oxford—I received word by letter from Fr. Fred Crowe, Lonergan's colleague and editor, that Lonergan had broken through the problem posed to him by the mess of theology and religious studies. Neither Fred nor I knew what the breakthrough was, but we had been hovering round the problem for a decade, especially in the context of Lonergan's search for the X called cosmopolis, some collaborative structuring of human loneliness that would make effective the drive towards authentic global happiness.[5] I plunged into the thesis in September of 1965 and indeed had a pretty complete manuscript by June of the next year. Meantime Lonergan had a lung removed in a cancer operation and was recovering in Regis College, Bayview Avenue, Toronto, and to that place I betook myself the following summer. It was then that Lonergan pointed me to his solution as we sat in his room, poised opposite one another as I asked the central question. He immediately swung his two hands towards each other, trembling fingers inches apart, and remarked laconically, "well it's easy: you just double the structure." He gave me a ten-minute lecture on the new heuristic of inquiry, a lecture that has kept me poised now for forty-seven years.

I returned to Oxford to push my thesis through. It was a matter, really, of hiring a typist to put the stuff in formal shape. My good and patient director, Rom Harré, was impressed by this move towards final shape. His optimism was all the more welcome in that I was in fact flying back and forth to Dublin in order to teach philosophy of science there. But the main point I wish to make is that I did not modify the first draft in the light of the radically new view of philosophy and of methodical collaboration. As it was, the thesis was peculiar enough. Rom and I mused over who might examine it and the final odd choice was a physicist and

---

[4] The specification was initially given in "Functional Specialties," *Gregorianum*, vol. 50, 1969, pp. 485–505. Later it was published as chapter 5 of *Method in Theology*, New York, Herder and Herder, 1972; vol. 14, Collected Works of Bernard Lonergan, Robert M. Doran and John D. Dadosky (eds.), Toronto, University of Toronto Press, 2017.

[5] The X of Cosmopolis was to be slowly identified, but the tragedy of its general reception by his followers is that they did not effectively recognize that there remained, scientifically, the X, the Higgs particle of religious studies predicted by Lonergan. It had been simply named in terms of old divisions, a shabby standard model, but what was/is X? That is a massively difficult challenge of serious understanding and heuristic envisagement. See Philip McShane, "Metaphysical Equivalence and Functional Specialization," *Method in Theology: Revisions and Implementations* (2007), http://www.philipmcshane.org/website-books, (accessed December 1, 2020).

a biologist, their names long forgotten by me. They were sympathetic readers but would not approve of the thesis without (1) my adding a clearer account of my method; (2) the elimination of chapter 8 of the book, which they considered to be merely pure mathematics.[6] It was at this stage that Lonergan wrote me a letter which included the advice, "give the fellow what he wants ... get the union card." I added the simpler account, thus replacing chapter one of the book in acceptable fashion, and I cut out chapter eight on "Foundations of Statistics." The revised effort is no doubt somewhere in the Bodleian Library, but the book published in 1970 and again now is the original manuscript.

What Lonergan did about his discovery of February 1965 is a story told elsewhere.[7] As I pondered over this introduction I returned to his post-*Method* writings, especially those contained in *A Third Collection*, musing over his strategy in—might I say?—hiding or underplaying his discovery. I like to think that he himself was taking his advice to me about giving the audience what it wanted or needed or could tolerate. In the essays of *A Third Collection* he rarely alludes to his massive shift in method, but the shift is echoed here and there.[8] Still, it is as well

---

[6] Was their claim true? The answer from a new culture of generalized empirical method is, No. "Generalized empirical method operates on a combination of both the data of sense and the data of consciousness: it does not treat of objects without taking into account the corresponding operations of the subject; it does not treat of the subject's operations without taking into account the corresponding objects." Bernard Lonergan, "Religious Knowledge," *A Third Collection*, edited by F.E. Crowe S.J., Paulist Press, 1985, p. 141; "Religious Knowledge," *A Third Collection*, vol. 16, Collected Works of Bernard Lonergan, Robert Doran and John Dadosky (eds.), Toronto, University of Toronto Press, 2017, p. 136. I refer to the book below simply as *A Third Collection* or CWL 16. This perspective underpins both the book and my main thesis here, about religious studies. Chapter 8 illustrates philosophy of mathematics in that new culture. An earlier version of this preface ventured into the manner in which various chapters impinged on philosophy in other domains, but—recalling note 3's point—it would carry us into other complex areas, such as contemporary refinements of randomness analysis in relation to computer vulnerabilities. But I cannot resist noting the relevance of chapter 4, "Reasonable Betting," to the undermining of various pseudo-statistics that parade as subtle wisdom in the destructive world of hedge funds.

[7] See Pierrot Lambert and Philip McShane, *Bernard Lonergan: His Life and Leading Ideas*, Vancouver, Axial Publishing, 2007, especially pp. 215–18, 236–39, 255–59. The book is referred to below as **Lonergan**.

[8] *A Third Collection*: "as yet the world religions do not share some common theology or style of religious thinking" (70; CWL 16, 68); "the cultivation of religious experience is its entry into harmony with the rest of one's symbolic system, and as symbolic systems vary with the culture and the civilization, so too does the cultivation of religious experience" (127; CWL 16, 123); "it envisages the conditions under which the study of religion and/or theology might become an academic subject of specialization and investigation" (129; CWL 16, 125); "In man, the symbolic animal, there is an all but endless plasticity that permits the whole of our bodily reality to be fine-tuned to the

to add a more fundamental view on this reticence expressed in Lonergan's answers to questions put by Professor Val Rice in those late interviews of the early 1980s:[9] he was leaving the push towards the meaning and the actuality of functional collaboration to his disciples.

The push towards the meaning of that collaboration has not found a serious place among his followers. Perhaps it is because, unlike Fr. Crowe and myself, they had not lived with the problem, the question, through the previous decade. Lonergan's most reliable presentation of his discovery appeared in 1969, prior to the International Florida Conference, but it was not aired in that conference, apart from my venture into its dire need in musicology.[10]

I must now go back to Oxford of the 1960s, and to the question of the sympathy of the readers and indeed the tolerance and patience of Rom Harré. I am going to be blunt and state that in that world, truncated subjectivity reigned supreme. "The neglected subject does not know himself. The truncated subject not only does not know himself but also is unaware of his ignorance and so, in one way or another, concludes that what he does not know does not exist."[11] Or

---

beck and call of symbolic constellations." (133; CWL 16, 128); "establish[ing] such a framework" (as against) "provide glib talkers" (136; CWL 16, 131). So he arrives at the issue of "generalized empirical method" (140–41, CWL 16, 135–36: see the quotation in note 6 above) and points the reader (note 7, 145, CWL 16, 136, note 19) to the beginning of *Method in Theology*.

[9] See **Lonergan**, pp. 110–12.

[10] The two papers that I presented at the Florida Conference—one on botany, one on musicology—"Insight and Emergence: Towards an Adequate Weltanschauung," and "Metamusic and Self-Meaning," appeared later as chapters 1 and 2 of *The Shaping of the Foundations*, University Press of America, 1976. A digitalized copy of this book is available at http://www.philipmcshane.org/website-books, (accessed December 1, 2020).

[11] Bernard Lonergan, "The Subject," *A Second Collection*, edited by William F.J. Ryan, S.J. and Bernard J. Tyrrell, S.J., London, Darton, Longman and Todd, 1974, p. 73; *A Second Collection*, Robert M. Doran and John D. Dadosky (eds.), vol. 13, Collected Works of Bernard Lonergan (University of Toronto Press, 2016), p. 64. Truncated subjectivity is a neurochemical patterning regularly associated with Scotus but it is comfortably Aristotelian and Thomist and increasingly inhabits Lonerganesque minds. It can rule the roost quietly in a discussion of questions, insights, desires: even laud generalized empirical method (see note 6 above) but never rise to theoretic reaching in that strange zone. Common sense comfortably detects patterns in commonsense talking and thinking, and even in the language of sciences, even rises to types of prescriptive metaphysics, but the serious climb of scientific understanding is only mimed. Truncated subjectivity certainly is not authentically, existentially, at ease in Augustine's shock regarding "a word brought forth when we utter what we know, a word that is before all sound, before all thought of sound" (see Lonergan, *Verbum: Word and Idea in Aquinas*, vol. 2, Collected Works of Bernard Lonergan, Frederick E. Crowe and Robert M.

perhaps I might say that, for the truncated subject, what is known by precise naming and description, especially rich comparative description, is thus sufficiently, even essentially, known.

Now I might well extend my suspicion regarding truncation to Europe and America of the 1960s—was it some parallel sentiment that fermented the 1968 student disturbances?—and so talk of, say, pragmatisms, existentialisms, phenomenologies; but that is a venture of future functional history, a plethora of books. So I cut back to a discomforting particularity, to Lonergan's following in Rome, where he was condemned to teach in 1953, and make the claim that there was no miracle there of self-discovery. "To say it all with the greatest brevity: one has not only to read *Insight* but also to discover oneself in oneself."[12] In the paragraph leading up to this conclusion Lonergan is quite blunt. "One has also to be familiar with theory and with technical language. One has to examine mathematics, and discover what is happening when one is learning it and, again, what was happening as it was being developed."[13] These are quite crazy demands for the philosophers and theologians, students and faculty, of the Rome from which Lonergan's sickness rescued him in 1965. What was more palatable for the community of students interested in Lonergan's obvious revolution was the interest he showed, in that decade, in more subtly suggestive writings about meaning. Add to that what is being increasingly identified as a lightweight grip on Lonergan's position on truth, reality, objectivity.[14] Such an ethos, laced with a comparative reaching into truncated cultures, dominated the First Lonergan International Conference in 1970. Such an ethos has, in the main, dominated Lonergan studies in the decades since.

I push further now to ask you to consider the strange question, whether the ethos dominated Lonergan in that decade. But it seems best to paint broader strokes that go back to what I might consider his last mad reach for serious

---

Doran (eds.), Toronto, University of Toronto Press, 1997, p. 7 n. 8) and so knows nothing of psychic skin's potential for luminosity, religious or noetic.

[12] Bernard Lonergan, *Method in Theology*, p. 260; CWL 14, 244.

[13] For Lonergan's fuller expression of this last demand, "what was happening as it was being developed," see McShane, *The Road to Religious Reality*, pp. 36–37. That demand, fully articulated in omnidisciplinary fashion, becomes the heuristics of a geohistorical genetics of human progress in serious understanding.

[14] The point is well made in particular by Mark Morelli's work. For a summary from him see Mark Morelli, "Lonergan's Debt to Hegel and the Appropriation of Critical Realism," *Meaning and History in Systematic Theology: Essays in Honor of Robert M. Doran S.J.*, edited by John Dadosky, Milwaukee, Marquette University Press, 2009. For a context from him see Mark D. Morelli, *At the Threshold of the Halfway House: A Study of Bernard Lonergan's Encounter with John Alexander Stewart*, Chestnut Hill, MA, The Lonergan Institute at Boston College, 2007.

intimations of the world of theory: his lectures on mathematical logic of 1957.[15] The 1958 lectures on *Insight* were a transition to the inadequate intimations necessitated by compact and summary presentations. *Insight* itself was already a doctrinal book:[16] but now Lonergan became his own Fontenelle,[17] and one finds that same bent in the summer presentations of the following years.[18] So I find a

---

[15] The lectures are presented in CWL 18, *Phenomenology and Logic*, edited by Philip McShane (see note 1 above). My introduction speaks of the genius of his effort, especially in his astonishing three-day journey through Jean Ladrière's classic work, *Les limitations internes des formalismes* (see CWL 18, xx–xxi). The question sessions show that his audience were not in his ballpark, nor indeed were the masters of logic he deals with: "my answer ... quite independent of these people" (CWL 18, 330) is a comment that could spread across the entire effort. Think of the Morelli context of the previous note and then ask, What might either audience make of the comment "the criterion of truth is the virtually unconditioned"? CWL 18, 338. But Lonergan knew that more precise control of the data was needed. In what was almost our last conversation in the early 1980s he shared with me his ongoing puzzling about the meaning of Gödel's Theorem. Later I gave a detailed invitation to the climb to its elementary meaning, relying on excellent work by Penrose, in "Gödel's Incompleteness Theorem," chapter 1 of Philip McShane, *Lonergan's Standard Model of Effective Global Inquiry* (2007), http://www.philipmcshane.org/website-books, (accessed December 1, 2020). Its full meaning is another matter, a matter of adding theorems of incompleteness to the simple presentation of "positioning" of *Insight: A Study of Human Understanding*, vol. 3, Collected Works of Bernard Lonergan, Frederick Crowe and Robert Doran (eds.), University of Toronto Press, 1992, 413, theorems reaching toward an eschatology of incompleteness correcting Paul's claim of completeness in *First Corinthians* 13:12.

[16] For decades I have been drawing a very enlightening paralleling of *Insight* with the classic graduate text, Georg Joos, *Theoretical Physics* (London, Blackie & Son, 1951), a book I used in 1956 as a graduate. I met *Insight* the following year: it is obviously a graduate text written boldly in the absence of undergraduate texts. So, Joos treats, in thirty pages, "the mechanics of systems of particles" (106–35), a topic that requires a set of undergraduate texts to understand. Think, now, of that great stumbling block of Lonergan studies, the thirty odd pages of chapter 17, section 3, on interpretation, on the mechanics of systems of persons (see note 25 below). My comments here obviously broaden the point I am making about the relevance of the struggle through *Randomness, Statistics, and Emergence*.

[17] Herbert Butterfield nicely identifies, in the final chapter of his *The Origins of Modern Science, 1300–1800* (New York: The Free Press, 1965), Fontenelle's role in *haute vulgarization*.

[18] I pause only over the final lecture of the 1959 lectures on education published as *Topics in Education*, vol. 10, Collected Works of Bernard Lonergan, Frederick E. Crowe and Robert M. Doran (eds.), University of Toronto Press, 1993. In a conversation with Lonergan about them at the time of their early editing he remarked, "I was just trying to work out a few things." In that final lecture, on history, he was brooding over "the problem of general history, which is the real catch." CWL 10, 236. A careful reading of the text can detect his struggle to give a positive lift to *haute vulgarization*. The solution to

sad irony in his comments on *haute vulgarization* of 1962[19] and 1963.[20] His audience was, and it seems to me it continues to be, largely an audience "with no real grasp of theory of any kind ... never bitten by theory,"[21] "lost in some no man's land between the world of theory and the world of common sense."[22]

It seems useful, if shocking, to return now to muse on some aspects of the volume already mentioned at notes 6 and 8 above, *A Third Collection*. First, lucky for us, the index is simply a proper-name index, with four pages of names. Many of these are familiar names in religious studies of the mid-twentieth century. Lonergan takes them respectfully, tolerantly, in their own contexts, but gives the odd nudge towards the need for a larger view. The nudges are modest. Nowhere does he insist openly on the need, either, for a context of serious theoretic understanding, and only occasionally does he point to the convenience of a functional division of labor.[23] Is he "giving the audience what they want," what they can tolerate? Neither his audience nor his cited authorities are beyond the world of Linnaean classification. But, you may gasp, do they need to be? I counter with the claim that a Linnaean classification of *Phaseolus limenses*, lima beans, is not contemporarily significant for fertilizing their growth; what of the cultivation of *genus humanum*, human beings?[24] My odd question leads us into the discomforting context of *Insight* chapter 17,[25] but here particularly to the section on "*The Genesis*

---

his problem was over five years away, but hidden in its expression in the meaning of the single word, "*Comparison.*" *Method in Theology*, p. 250; CWL 14, 235.

[19] "Time and Meaning," reproduced in CWL 6: see page 121.

[20] "Exegesis and Dogma," CWL 6: see page 155.

[21] "Exegesis and Dogma," CWL 6, 155.

[22] "Time and Meaning," CWL 6, 121.

[23] See *A Third Collection*, p. 179, with the corresponding note 13 on page 183 (CWL 16, 173, note 15), and p. 196 (CWL 16, 191–192), where he points to the key role of the two specialties, Dialectic and Foundations.

[24] It is as well to recall Lonergan's broad pointing to the issue: "The initiative seemed permanently in the hands of those who invoked science against religion, and it mattered little to them that at any given moment the issue had shifted from physics to Semitic literature, from Semitic literature to biology, from biology to economics, or from economics to depth psychology." *Insight*, CWL 3, 755.

[25] I refer especially to the third section of the chapter, where Lonergan takes off in the crazy solitude of the summer of 1953 to talk of theoretic eigenfunctions of meaning, something he only accurately determined in 1965 as eight in number. With that eightfold way there can be foreseen, at present only in fantasy, the "fuse into a single explanation" that ends the climb of the second paragraph of the second canon of hermeneutics. *Insight*, 610, line 9. But let me add concrete embarrassment to my sweeping words. The volume titled *Lonergan's Hermeneutics: Its Development and Application*, (Sean McEvenue and Ben Meyer (eds.), the Catholic University of America Press, Washington D.C., 1989) was the result of our (yes, guilty!) gathering on that topic

of Adequate Self-Knowledge," with its demand for education in the simple science of physics and the blunt conclusion of its second paragraph: "Most of all, what is lacking is knowledge of all that is lacking, and only gradually is that knowledge acquired."[26]

So I arrive at the place of such works as *Randomness, Statistics and Emergence* in the gradual acquisition of an aesthetic-toned[27] knowledge of our 21st century's need to start over. But it is not a simple placing, for our context is that described in chapter 7 of *Insight*'s account of bias and in his global solution to that sick living. Both the study of religion and the study of Lonergan are caught in complexes of biases, swept into a common comfort of general bias: they need the radical remedy of a glocal rinse cycle.[28]

This is a fiercely discomforting claim based on sixty years of journeying through the brutalities of institutions of the second half of the 20th century. It is expressed in my writings of the years between *Randomness, Statistics, and Emergence*, and *The Road to Religious Reality*. Perhaps it is sufficient in this short Preface to point you to the website project of the past ten years, the ten-volume Cantowers that weave together the challenge of *Insight* and *Method in Theology*. The particular challenge of scientific seriousness is bluntly expressed in chapter ten of the Lonergan biography, which insists that the context of progress cherished by Lonergan is that of competence in the simplest science. I would add here the deeper issue of competence in the more difficult next science, chemistry. That deeper issue is brought out, or in, only by a slow neurochemical ingesting—here the topic of the character of adult growth emerges as a crisis in human history[29]— of a self-luminous poising of Gown that would mediate a new Town psyche. The topic is too vast and novel to enter into here. But perhaps a stretch towards its fantasy in terms of randomness, statistics, and emergence would help seed hopes and focusing efforts. There is the randomness of present discontent with disorientations of religiosities and confusions of religious meanings, with corrupt banking and its idiot economic backing, with destructive eco-behavior and its

---

without adverting to having ingloriously missed the pointing. The point continues to be missed, or dodged.

[26] *Insight*, CWL 3, 559.

[27] See Philip McShane, "Aesthetic Loneliness as the Heart of Science," *Journal of Macrodynamic Analysis* 6 (2011), 51–84.

[28] The failure of Lonergan studies is more obvious in that there is an acknowledgment within the school that Lonergan has raised the bar without a measuring up to that raising (see, in particular, notes 11, 15 and 27 above). A sad consequence is the nonoccurrence of Lonergan's view or even name in standard present survey work, such as Ian Barbour, *When Science Meets Religion*, New York, Harper, 2000. Barbour's effort remains, then, extremely Linnaean.

[29] The crisis is briefly described on pages 161–63 of Philip McShane, *Lack in the Beingstalk*, Nova Scotia, Axial Publishing, 2006.

grounding in sick patterns of greed, with hidebound bureaucracies and their stranglehold on government. That randomness can shape up, in tiny collaborative steps, to a statistics of the emergent recurrence-schemes internal to a global Tower of Care.

> The probability of the single events are the same as before, but we cannot suppose that the probability of the combination of all events in the set is the same as before. As is easily to be seen, the concrete possibility of a scheme beginning to function shifts the probability of the combination from the product pqr ... to the sum $p + q + r + $ ... For in virtue of the scheme, it now is true that A and B and C and ... will occur, if either A or B or C or ... occurs; and by a general rule of probability theory, the probability of a set of alternatives is equal to the sum of the probabilities of the alternatives.[30]

The cyclically-summed actualities can, over millennia, shift from Poisson distribution to a Normal and normative law, giving supreme plausibility to a Tower of Able[31] of serious intimate[32] understanding grounding, literally, a plain plane of radiant life in the next million years.

---

[30] *Insight*, CWL 3, 144.

[31] See **Lonergan**, p. 163, "The Tower of Able: Lonergan's Dream," the concluding image of the book's Second Part, "Images of Lonergan."

[32] The intimacy is a matter of a shared inner word, "*eo magis unum*" (see the final chapter of Lonergan, *Verbum: Word and Idea in Aquinas*, CWL 2, 204–08) with a shared neurochemistry of imaged psychic tonalities. For Christian Tower-dwellers, that shared neurochemistry centres on the noise *Jesus* expressing the gradual ever-incomplete achievement of a field view of history. As a lead-in, see Philip McShane, "The Hypothesis of a Non-Accidental Human Participation in the Divine Active Spiration," *METHOD: Journal of Lonergan Studies N.S.*, vol. 2, no. 2, 2011, pp. 187–202. Other religious groups must find their way to their specification of the sequence of meanings of history lurking in the heuristic word *Comparison* on page 250 of *Method in Theology* (CWL 14, 235). But of course here I have been writing of the Tower People. The plain plane people move differently, tower-mediated in the "Yes," "and" of Faith, Molly's "yes" that ends *Ulysses'* each Bloomsday day's riverrun, and a Finneganend "and"— where Finnegan beginsagain, opening loneliness to tomorrow and eternity.

Randomness, Statistics, and Emergence

This volume contains an attempt to do four things at once, and this fourfold attempt influences the structure and the content of the work in various ways.

First of all it is an attempt at inaugurating dialogue between various schools of philosophy. Because of this, the work will evidently be difficult reading for the members of any one school. It was written at Oxford, and draws on that background, yet it is not of Oxford, for its philosophic stance is continuous with the structured critical realism of Bernard Lonergan. Yet the writing was governed by the appreciation of the chasm between the two views of philosophy, and so the philosophic position presupposed by the entire work becomes explicit only in the concluding chapter. It is, as the reader will see, drawn out of the content of the work itself: it emerges primarily within the discussion of emergence as closely related to the prime analogue for the understanding of emergence.

Secondly, the work is an attempt to orientate philosophy of science away from general considerations towards a reflection on the detailed content of science and on the details of procedure in scientific investigation. So, for example, in chapter eight, instead of dealing in general terms or in terms of philosophic positions with the problems of the foundations of statistics, we deal with the content and expression of some standard work in probability theory. For, the philosopher of science has the task of transforming and correctly orientating science from within, and he can do so adequately only in so far as he includes as data for philosophic reflection the details of contemporary science.

Thirdly, the work is intended for those who already share the author's basic philosophic position. It offers, then, a more detailed account of various points treated in the works of Lonergan, particularly of course the principle of emergent probability, but also more precise indications, such as a possible meaning of the change from the first to the second edition of *Insight* on the question of convergence in probability sequences. Also, throughout the book are scattered hints which might be followed up: regarding the foundations of Geometry and Space-Time structure in chapter six, regarding metalogic,[1] regarding the detailed meaning of randomness and coincidental aggregates, regarding the complementarity of classical and statistical science and the significance of recurrence-schemes, etc. These hints are given, positively in the hope of expansion, negatively to counter the possibility of the contraction of decadence. Any worthwhile philosophic position is open to the possibility of such contraction, and a standard manifestation of such contraction is the summary.

---

[1] Cf. also the chapter 'Symbols and Syllogisms,' *Towards Self-Meaning*, G. Barden and P. McShane, New York, Herder and Herder, 1969.

The summary can give the impression of capturing the essence of a position. But a summary expresses the essence only in so far as the summarizer has the essence of the position in his mind. In this respect one may note that the book *Insight* is a summary expression of a philosophic position. As such it provides a phantasm for the reader which requires elaborate supplementation if the reader is to reach the mind of the author. The present work, it is hoped, is a contribution to that supplementation. Obviously, were this supplement written merely for the members of one school, it would have taken a more straightforward form such as I have attempted elsewhere.

Lastly, then, there is the central positive contribution of the thesis. It is an attempt to establish on a wider basis of contemporary mathematics and science the position of B. Lonergan on the nature of randomness, statistics, and emergence. Not only might that wider basis, as I remarked, have been given in more straightforward fashion, but it is given only partially, and this again in the interest of dialogue. So, for example, the notion of central form is consistently passed over: its exclusion will be evident to those familiar with Lonergan's position; its inclusion would have been a stumbling block to those of other schools.

At one stage in the present work I considered the possibility of including a certain degree of dialogue with phenomenological thought, but such an inclusion would have complicated the conversation overmuch. Suffice it to say that what is expressed in the work springs from a critical phenomenology of the scientific mind.

The book might well have been subtitled *Towards an Adequate Weltanschauung.* Such a subtitle may appear somewhat extravagant since the topics dealt with seem to be special questions in the philosophy of science. As I have argued elsewhere, however, in so far as one is a serious thinker, claiming an adequate viewpoint, a central element in that viewpoint is one's thought on the relationship of chemistry to botany.[2] Without that thought one lacks a basic component for the conception of world process. The present work deals with the central element and the heuristic conception of world process. It tries to lead the reader towards an adequate *Weltanschauung* through a dialectic of personal performance. The world view that should emerge—in the reading subject—is the world view of *Critical Existentialism,* a view which originated with Bernard Lonergan. Critical Existentialism contrasts with other forms of existentialism in that it seeks to mediate the subject's authenticity through a self-appreciation of metascientific dimensions. Again, the present work seeks only to lead the reader critically *towards*

---

[2] 'Image and Emergence: Towards an Adequate *Weltanschauung*' is a paper that was read at the First Lonergan Congress, Florida, 1970; later published in *The Shaping of the Foundations: Being at home in the Transcendental Method*, ed. Philip McShane, Washington, University Press of America, 1976, 6–45.

an adequate world view. I have tried elsewhere, in a symbolic personal pattern, to indicate further fundamental components of a *Weltanschauung* which would be adequate to our times.[3]

For the reader familiar with B. Lonergan's *Insight* it may be useful to note certain parallel discussions. There is first of all a rough correspondence between chapter two of the present work and the characterization of non-systematic process in *Insight* (pp. 48 ff; CWL 3, 71 ff) and its association with the inverse insight of statistics through the need to consider events as members of a group in order to reach a fully general account of the meaning of random differences (*Insight*, p. 55; CWL 3, 78). Secondly there is a parallel between the third chapter of the present work and the argument for the existence of statistical residues summarized in section 6.5 of *Insight*. Again, while chapter six of *Randomness, Statistics, and Emergence* touches on a large range of problems, the movement of the chapter is towards an appreciation of the definition of chance suggested in *Insight* (pp. 114, 125, 132; CWL 3, 137, 148, 155). Other parallels are indicated above or in the text, or are already sufficiently evident.

Thanks is due to a large range of persons but we must restrict ourselves here to those more immediately concerned: to Rom Harré of Oxford, who supervised a version of the present work as a thesis; to Garrett Barden and the Rev. Conn O'Donovan S.J., both of the Milltown Institute of Philosophy and Theology, Dublin, for their encouragement, discussions, and suggestions. There is, finally, the more obvious debt to Rev. Bernard Lonergan S.J., Regis College, Toronto.

---

[3] *Music That Is Soundless*, Dublin, Milltown Institute, 1969.

Randomness, Statistics, and Emergence

1 PROBLEMS AND CONTENT AND THE PROBLEM OF METHOD

While the short title of the present work, *Randomness, Statistics, and Emergence*, explicitly mentions the key elements in the content of the thesis, the thesis title, "The concrete logic of discovery of Statistical Science, with special reference to problems of evolution theory," is more explicit about the general underlying problem. That general problem concerns the logic of discovery, and the entire work may be regarded as an experiment in that logic, in methodology, heuristics, in prescriptive metaphysics. The logic is named 'concrete' in the title rather for emphasis than of necessity, for an adequate logic of discovery must necessarily be concrete, where 'concrete' connotes not only a close attention to the empirical but also a total anticipatory inclusiveness. First, the empirical attended to is not just the empirical of common sense or science but the empirical which is actual scientific practice in the process of discovery. Secondly, the inclusiveness should be both total and anticipatory. The total inclusiveness demands that in some way and to some extent the findings of the logic of discovery should lie outside the domain of the conceptual schemes of science. Such conceptual schemes do in fact anticipate in their own fashion a filling-out by further investigation: but they are essentially replaceable.[1] On the other hand one would hope that some core of the logic of discovery would be invariant. Nor is this hope unfounded: for, the replacement or revision of conceptual schemes presupposes some principle of revision which persists and carries through from one scheme to the next. In so far as it does thus carry through one may say that it is total and invariant: for it must include in some way the total sequence of revisions without being subject to basic revision itself. Furthermore, one can specify the inclusiveness as anticipatory. 'Anticipation' here means something different from the presence of conclusions in premises. What is meant is that type of anticipation one has, for example, when one specifies science as theories verified in instances. The practising scientists can recognize in that brief statement a formula for an acceptable scientific paper, but the formulation of the theory, the particular instances and samples, and the type of verification will all differ from paper to paper, from science to science. The formula, 'theory verified in instances,' anticipates heuristically, and it is filled by scientific work. Finally we may note that the brief characterization of science as theories verified in instances seems itself to be a clear illustration of an invariant of the logic of discovery.

But already we are anticipating our own general conclusions regarding the logic of discovery when in fact the very existence of such a logic is a matter of dispute. In a recent discussion of that dispute N. R. Hanson remarks that

---

[1] Cf. Part 1 of R. Harré, *Matter and Method*, London, Macmillan, 1964, on conceptual systems in general; as a determinant of method, cf. ibid., 108ff.

as an area of inquiry which, in my opinion, has received far too little examination by logicians and philosophers, the context of Discovery should be recognized as having logical credentials of its own and should not be relegated to being a kind of "puzzle out in reverse" of what will end up as the finished research report.[2]

Hanson's approach to this question is, however, more the exception than the rule. Many others reject the possibility of a logic of discovery for a variety of reasons. A general reason for this rejection comes from conceiving logic uniquely as deductive logic, and to reject logic of discovery in this sense is no great achievement:

> When Popper, Reichenbach and Braithwaite urge that there is no logical analysis appropriate to the psychological complex which attends the conceiving of a new idea, they are saying nothing which Aristotle or Peirce would reject. The latter do not think themselves to be writing manuals to help scientists make discoveries. There could be no such manual.[3]

A refinement of that reason for rejecting the logic of discovery would point out that past discoveries or present data do not determine future hypotheses: these hypotheses emerge, therefore, as indeterminate conjectures, and one cannot have rules for determining what are essentially indeterminate. A further refinement would point to the fact that the data of that logic of discovery would have to be in some way the actual emergence of the discovery, the hypothesis, in the discoverer's mind, and that such data elude scientific investigation, or, more radically, are to be excluded as lacking scientific respectability.

However, there is no denying the growing interest in the processes of discovery, learning, development of understanding etc., both within the field of education, psychology and psychiatry and within the field of history of science and scientific method. Studies in these various regions reveal that the growth of understanding, either in the individual or in the group, is far from totally haphazard. In so far as it is not, one may expect it to have some structure. If that structure can be appreciated—and this is the problem which will occupy us in chapter twelve—then it can also be to some extent formulated. For this reason, Hanson's remark, that the logic of discovery could not find expression in a manual, would hold only in a strict sense. For one thing, manuals on method do exist, and the methods relate to the generation of understanding in oneself and others. Certainly a manual on method or on the logic of discovery cannot consist of a list of axioms from which discoveries follow. But besides axioms which virtually contain unknown theorems there are canons which can heuristically

---

[2] N. R. Hanson, 'The Idea of a Logic of Discovery,' *Dialogue*, IV, 1965–6, 60–61.
[3] Ibid., 49.

contain the unknown. Here one is reminded of Mill's four canons,[4] or of Lonergan's six canons of empirical method.[5]

These canons and the associated logic of discovery require for their adequate contemporary formulation a close attention to the total field of modern science. In his *A Hundred Years of Philosophy* Passmore remarks that 'until about 1960, philosophy of science and history of science were sharply sundered and most philosophers of science were content to illustrate their reflections on scientific methodology by, at most, one or two stock examples.'[6] The situation has changed considerably in the past decade and now much more attention is being paid to actual scientific achievements of the past and present. Still, there is an ever-present tendency in philosophy of science to limit considerations to the general features of scientific inquiry and scientific theories and thus to become remote from the actualities of scientific practice. In the present investigation that tendency is countered by constant recourse to the wealth of contemporary scientific endeavour.

A point on which we will insist throughout is the *a posteriori* nature of the canons of method and of the logic of discovery in general. This too is a basic reason for postponing the effort to characterize the logic of discovery to the end of the present work. Scientific method and the possibility of a logic of discovery can only emerge and be appreciated through the development of science and the growing interest in method. The word 'emerge' is used here advertently and the discussion of the meaning of that word in chapter nine will in fact put us in a better position to discuss, in the concluding chapter, the meaning and emergence of the logic of discovery.

Leaving aside the question of method, we turn now to a brief consideration of content. At first sight the problems to be dealt with in the investigation would seem to lack unity. There would appear to be present the clear duality of a discussion of statistics and another discussion of evolution or emergence. But the duality fades when one comes to appreciate that the key to the two sets of problems, those of the philosophy of evolution theory and those of the philosophy of statistics, lies in the analysis of randomness and that randomness has, so to speak, two faces. With each of the two topics, statistics and emergence, we associate eight problems, and we list immediately these two groups of problems in a manner which brings out their isomorphism: Ai and Bi correspond.

List A: Statistics.

A1. Are statistics a cloak of ignorance?
A2. What is the 'object' of statistics, the *explanandum*?

---

[4] *A System of Logic*, London, Harper Brothers, 1896, Bk. Ill, ch. 8 ff.

[5] *Insight: A Study of Human Understanding*, London, 1957; CWL 3, ch. 3.

[6] J. Passmore, *A Hundred Years of Philosophy*, London, Duckworth, 1966, 541, footnote.

A3. What are the units of definition in statistics?
A4. How does statistical explanation relate to causal explanation?
A5. Do statistics apply to the particular case?
A6. What is wrong with the limit definition of probability?
A7. The paradox of Laws of Chance, chance being the exclusion of law.
A8. How are statistical hypotheses verified?

List B: Emergence.

B1. Is emergence an epiphenomenon?
B2. How is the 'object' of evolution theory specified?
B3. What are the units of evolution in evolution theory?
B4. How combine here statistical and causal explanation?
B5. Does evolution theory involve a statistics of the particular case?
B6. What is wrong with contemporary statistics of evolution?
B7. The paradox of emergence of system on a background of chance.
B8. Can an evolutionary hypothesis be verified?

The solution of these sixteen problems forms a unit and the partial solutions come out to some extent in pairs because of the fact of randomness and its 'two faces.' One face of randomness looks, so to speak, towards statistics: statistical science is found to provide a type of general knowledge of random aggregates. The other face of randomness is towards emergence, where there is a connotation of another mode of systematizing randomness on various levels. The focus of attention in the thesis is, then, randomness on various levels, its occurrence, its types, its general knowledge and systematization, its significance in evolution theory.

The first stage of the treatment will centre on the occurrence and meaning of randomness on the level of mathematics. Randomness on this level is appreciated elementarily by considering the difference between such simple series as the following:

1 2 3 4 5 6 7 8 . . . .
6 7 5 6 6 7 6 5 7 6 5 6 . . . .
5 1 8 4 3 9 2 3 7 4 9 6

One understands immediately—and this is indicated by the dots at the end—how the first series would be continued: the series is systematic, with a formula for the nth term. The second series, too, ends with a sequence of dots, but these do not indicate the existence of a formula. But while there is no formula one nonetheless reaches some understanding of it in so far as one grasps that the series oscillates about 6 in a way which may be called random. Here we have the type of series which occurs in statistics. The third series does not end with dots: there are no rules whatever for its continuation. In so far as there is no law relating to it, it may be described as totally random. Randomness, then, may be identified with some type of irregularity, an irregularity which excludes system but which may at times be to some degree specified. A discussion of series of the second type will lead us to an elementary account of the basis of statistics.

The next step is to move, in chapter three, from mathematics to physics, thus raising the question, whether randomness occurs in nature, whether system is anywhere objectively excluded. One can show, for example, that there is no formula for the locations of the people standing round at a cocktail party, but one must show further that the set of locations cannot be systematically deduced. This can be done by showing that the set of preconditions for any one location form a non-systematic aggregate and that in general the number of conditions diverge as one moves to prior conditions. One cannot therefore move back to a unified set of conditions for all the locations. This does not make the locations or the antecedents indeterminate: they are determinate, but non-systematic. Laplace's view of the deduction of states of the universe can be considered at this stage, and its weakness shown to depend on his neglect of the non-systematic. In this context also Aristotle's failure to reach statistical understanding of the random or non-systematic is pinpointed. Aristotle considered the non-systematic to be outside science and associated with it the notion of chance. A precise consideration of chance is given in chapter six.

In chapter four a clearer appreciation of the method of statistical science is reached by considering the relation of the statistical hypothesis to the single instance: it is shown that the relation has the character of reasonable betting. This consideration clears the ground for a discussion in the following chapter of the manner in which statistical inquiry is related to causal or classical inquiry. Statistical science depends on causal inquiry for its definitions, and it complements causal inquiry in many ways in reducing the residue of unexplained phenomena. But the existence of random situations excludes the possibility of eliminating the statistical residue, real aggregates which exclude system and of which general knowledge is obtained through statistics.

Chance, like randomness, has been considered to involve a certain exclusion of rule. The relation of chance to randomness is specified in chapter six, this leading to a resolution of the paradox expressed in the phrase, 'Laws of Chance.' The Aristotelian definition of chance is replaced by a definition which links chance to the random fluctuations from the ideal norm given by a statistical theory, for example, the fluctuations from 6 in the second series, page 5 above.

In chapter seven a consideration of certain peculiarities of the procedure of verification of statistical hypotheses leads to some clarification of the process of verification in general. Here too the basis of the distinction between the two concepts of probability is exposed by reference to the Aristotelian division of questions. In this, as in earlier chapters, points relevant to the problem of the formulation and verification of an evolutionary hypothesis are noted. A further relevant point is made in chapter eight in the discussion of more basic difficulties in the definition of empirical probability. Here a certain tendency to abstractly oversimplify is noted and the need for the development of more refined and empirically-referent probability theories is revealed. This tendency to depart from the empirical is a characteristic of discussion of evolution theory not only within the philosophy of science but also within science itself, and so, for example, the principle of evolution tends to be a general abstract statement rather than an open and up-to-date methodological statement.

Before dealing with the evolutionary hypothesis it is necessary to investigate fully the second aspect of randomness. This is done in the long chapter entitled 'Randomness and Emergence.' There it is shown that, for example, the phenomena called the amoeba cannot be accounted for systematically either through biophysics or biochemistry. On either of these levels the amoeba-reactions can be explained only non-systematically by an aggregate of formulae and equations. Moreover, the non-systematic account for any one amoeba is found to be non-generalizable, not applicable to another amoeba. This chapter is especially illustrative of the empirical attitude in the philosophy of science. The dominant attitude throughout is 'Try it and see,' and so, for example, the possibility of biophysical systematization is rejected not *a priori* but by drawing on the underlying principles and evident conclusions of the works of N. Rashevsky and of the standard journals devoted to biophysics. In this way randomness is shown to be the basic condition of the emergence of different levels of things, and the justification for the existence of different irreducible sciences.

One further clarification is required before focusing our results regarding statistics and emergence on the problem of a methodological account of the evolutionary theory. That clarification concerns the vexed question of the units of evolution. Here again the answer comes from closer consideration, in chapter ten, of the work of a range of scientists, this illustrating once more the need for close attention to present scientific practice and results in order to arrive at new results in the philosophy of science. The relevant units are found to be, not individual organisms, nor populations, nor genes, nor gene pools, but cyclic schemes such as Krebs cycle or the nitrogen cycle, cycles which may include, or be included in, individual organisms.

The second last chapter brings the previous elements together to reach a formulation of an adequate methodological principle of evolution. The principle is seen to be methodological in a strict sense: it gives a structured anticipation of the type of statistico-causal explanation that can be expected from the collaboration of sciences. It anticipates the best that this collaboration can do, which is, to give an account of the actual process of evolution and of emergence in terms of successive schedules of empirical probabilities of the series of conditioned recurrence-schemes that have actually emerged. This best is seen in anticipation to be a determinate but non-systematic account: in being non-systematic it reveals the flaw in Laplace's view; in being determinate it excludes an indeterminist view.

As we posed our problem originally in terms of sixteen related questions, so here we may bring together the particular results in the sixteen corresponding answers.

List A: Statistics.

A1. Statistical science is not a cloak of ignorance: randomness is objectively given.

A2. Its *explanandum* is formulated in an axiom of randomness.

A3. Its units of definition come from the classical, causal definitions of the aggregate-elements.

A4. Statistical explanation is complementary to causal explanation and to classical method.

A5. Statistical method does not deal with the particular case.

A6. The definition of empirical probability at present is not adequate, nor should it be unique: it should depend on the randomness involved, thus taking more account of the empirical.

A7. Chance can be more precisely defined (than the accepted Aristotelian sense) in terms of random divergences from probabilities.

A8. The mode of verification of statistical hypotheses differs from that of causal hypotheses, involving intelligent sampling etc., in a manner which throws light on the general problem of verification.

List B: Emergence.

B1. Emergence is not an epiphenomenon.

B2. The *explanandum* of evolution theory is the partly random process of emergence as it actually occurred.

B3. The units of evolution theory are 'concretely-operable' classical schemes.

B4. The units are built into the statistics, thus combining statistical, systematic and historical explanation in methodological fashion.

B5. No 'statistics of the particular case' are involved.

B6. The flaw in much of contemporary evolutionary statistics and discussion is an absence of empirical reference, both in its units and in its statistical methods.

B7. Emergence and evolution are explained in terms of probabilities of emergence and probabilities of survival of recurrence-schemes, with transitions—based on random mutations and interactions—from non-systematic to systematic occurrence.

B8. The mode of verification of the principle of evolution as here formulated has certain peculiarities. The principle is methodological, and in so far as it is founded on basic patterns of classical and statistical investigation it is tested and revised only in its content.

We have surveyed quickly the content of the next ten chapters of the present work here not only to facilitate its reading but also to raise the questions which will be considered in the concluding chapter and, in this way, to make possible a change in the focus of attention in that reading. Our survey shows that we hope to reach a coherent set of conclusions about the nature of scientific investigation and the structure of the object of that investigation. As was to be expected from the introductory remarks, these conclusions are heuristic and concrete rather than abstract. Moreover, while the conclusions form a coherent set, some of these conclusions lie closer to the invariants of scientific investigation—as, for example, random aggregates as basic to the distinction between sciences—while others, such as the suggestions about types of statistics in chapter eight, are more tentative. Again, the argument, even for those conclusions which we would consider to lie close to the invariants of scientific method and the logic of discovery, is cumulative. Thus the argument for real randomness is nowhere presented in a syllogism: just as the evidence for it lies scattered throughout the sciences so the elements of the argument for it lie scattered through the chapters, from the mathematics of chapter one to the necessity of including a statistical component in our best effort to reach an explanation of the universe in chapter eleven.

But the basic question which we wish to raise now only to postpone its consideration to chapter twelve is, What sort of conclusions are these and how are they reached? While our interest is in the logic of discovery, the conclusions reached seem to resemble to a great extent what may be regarded as metaphysical propositions, conclusions about the structure of reality. How precisely are our conclusions related to metaphysics? By what type of philosophical reflection are they reached? These questions would seem at first sight to lie at best only on the fringe of the central topics of randomness, statistics and emergence, but we will see in the concluding chapter that the two types of question, of method and of content, are intimately related through the paradigm for the understanding of emergency. That relation is such that if it is missed, our conclusions regarding real randomness, the objectivity of statistical science, and the fact of emergence, are liable to be misconceived. Hence the reflection on method reveals itself to be essential to the understanding of the content.

In an article entitled *Analyse, Science et Métaphysique*,[7] Strawson distinguishes various modes of philosophical reflection. A basic distinction is between those who seek in philosophy some type of formal system and those whose philosophic efforts consist in informal but critical attention to ordinary discourse. Strawson pinpoints clearly the inadequacies of the formalist view and goes on to characterize five modes of philosophical inquiry which are not exclusively formal or informal. Thus, one may centre one's attention on resolving paradoxes, or one

---

[7] In *La Philosophie Analytique*, Paris, 1962.

may widen the scope to include an investigation of why our concepts function as they do, further even, to ask how things would appear if we were different. Again, in a fourth mode, one may pursue analysis in a nontherapeutic fashion, although such analysis may well have a therapeutic result.

In characterizing a fifth mode, Strawson makes his well-known distinction between revisionary and descriptive metaphysics, the latter pursuit not only giving rise to particular clarifications but 'aiming to lay bare the most general features of our conceptual structures.'[8] This latter aim of philosophical inquiry is clearly close to our own, and Strawson's concern with 'categories and concepts which, in their most fundamental character, change not at all'[9] finds an echo in our interest in invariants. Moreover, our procedure, like Strawson's, seems to be more by way of ordinary discourse than through formalization. But our discussion moves more easily into the realms of science than Strawson's does, and related to this ease with science there is a stress on metaphysics as *prescriptive,* for, like R. Harré, 'it is my belief that useful and respectable prescriptive metaphysics arises naturally in the discussion of the sciences.'[10] The resulting prescriptive, even revisionary, stress, however, is not such as to put our results in Strawson's class of revisionary metaphysics, for our conclusions seem in fact remarkably close to the claims of common sense. The revision involved, indeed, is not so much one of common sense as one of philosophic method and scientific presuppositions. And this point brings us close to the basic difference between our prescriptive and Strawson's descriptive metaphysics: for that difference is most clearly evident in the association of prescriptive metaphysics with a methodology or a logic of discovery. That association reveals that the centre of interest and attention in prescriptive metaphysics is somehow the preconceptual rather than the conceptual or the field of conceptual analysis. The association remains to be explained and justified in chapter twelve.

To return to the initial statements of this chapter, the basic question about a logic of discovery is the question of its possibility and the possible method of pursuing it with success. That question, according to our general *a posteriori* principle, is to be settled by the fact of a successful performance. So, we prefer to postpone discussing the question of method further till we have successfully engaged in what we would consider to be the relevant method. The intervening chapters, however, are not written with a stress on the underlying method of philosophical inquiry. For this reason, these chapters may be read most easily with an interest in the particular results, and this either from an ordinary-language point of view or from the point of view of the scientist. But they may be read more adequately with an interest in the implicit philosophical method.

---

[8] *Individuals: An Essay in Descriptive Metaphysics*, London, Methuen, 1959, 9.

[9] Ibid., 10.

[10] *Theories and Things*, London, Sheed & Ward, 1961, 2.

The argument, then, runs on two main levels. Firstly, it moves cumulatively towards the particular set of results. Secondly it moves, more implicitly but still cumulatively, towards a justification of the method by which the results were reached. To the problem of making precise the nature of this method, and the nature of the particular results in the light of that precise specification of method, we will return in the concluding chapter. But it is as well to note here that there is presupposed throughout a view on objectivity which is of a piece with the philosophic method. It is of a piece with the philosophic method, which is one of attention to oneself in the performance of knowing, because it can only be established thematically through the subject's attention to his or her own performance in knowing. Briefly, the basic assumption is that the real is what is to be reached by correct understanding. That assumption cannot be intelligently challenged without being performatively contradicted in the process of challenging: it is the basic assumption of all intelligent dialogue. In so far as it is not accepted, then many of the arguments of later chapters will appear inconclusive. So, for example, one might agree that it had been proved that randomness is a necessary feature of the world-as-correctly-understood, yet still raise the question of the world-in-itself. But what is a necessary feature of the world-as-correctly-understood is, according to the assumption, a necessary feature of the world-in-itself. The assumption, then, is obviously central to the meaning of the thesis: it will be treated in brief in chapter seven and more fully in the context of the discussion of results and method in chapter twelve.

Normally books on probability and statistics begin with some short discussion of the fundamental notions of statistical theory. Thus Kendal and Stuart[1] in a very brief space point out that 'the fundamental notion of statistical theory is that of the group or aggregate ... called a population,' and go on to elaborate a definition of statistics as 'the branch of scientific method which deals with data obtained by counting or measuring the properties of populations of natural phenomena.' Yule and Kendal[2] further point out that statistics deals 'with data effected by numerous causes' and that 'by statistical methods we mean methods specially adapted to the elucidation of quantitative data effected by a multiplicity of causes.' Few authors of such treatises take it upon themselves to investigate fully precisely how aggregates form the object of statistical investigations, precisely how the data are effected by a multiplicity of causes. In dealing with the former question of how aggregates provoke statistical investigation the notion of randomness is inevitably referred to, and various authors, like von Mises,[3] try to formulate a basic assumption of randomness. Again, we find in Arley and Buch[4] a list of reasons for statistical investigations which touch directly on the second question above, the question of causes. The reasons are worth noting somewhat fully.

The first reason they give is the difficulty of defining an initial state accurately enough to be able to determine the final state uniquely. This is the case where a very small variation in the initial state can cause a large variation in the final state, even though the phenomenon be simple in character. For example, an extremely small change in the initial velocity of a roulette wheel can decide whether the final state is black or red. There is also the familiar example of molecules in a gas. Again, the initial state may even be unmeasurable in principle because a measurement would produce an uncontrollable change in the phenomenon investigated.

The *second* reason for the random character of a phenomenon may be related to the laws of nature involved. This is the case:

(a) If the relevant laws of nature are so complicated that in practice it is impossible to make the theoretically possible calculation of the final state. An example of this is a series of throws with a die; only a few throws show that the results are randomly distributed. Theoretically one could predict the result of a given throw if one knew

[1] *The Advanced Theory of Statistics*, London, Hafner, 1963, Vol. I, 1, 2.

[2] *Introduction to the Theory of Statistics*, London, McGraw-Hill, 1950, xv, xvi.

[3] R. von Mises, *Mathematical Theory of Probability and Statistics*, New York, Academic Press, 1964, edited and complemented by Hilda Geiringer, 6.

[4] *Introduction to the Theory of Probability and Statistics*, New York, Wiley, 1950, 3–5.

the exact initial position and velocity of the die, its geometric form, its mass, its moment of inertia, the elastic properties of the die and of the table, and so on. Thus an extremely accurate analysis of all these cooperating causes would be necessary, but such an analysis would be so complicated that in practice it would be impossible.

(b) If the relevant laws of nature are not sufficiently well known; this is true, e.g., in most biological phenomena. Thus if we measure the heights of a number of soldiers, the results are conditioned by a number of biological processes such as heredity and nutrition, the regularities of which are only partly known.

The *third* reason for the random character of a phenomenon is that all laws of nature are strictly valid only for *idealized* phenomena ... In reality the phenomena are always complicated and, furthermore, are subjected to disturbing factors such as changes in temperature and pressure, shocks, ... Such disturbing factors are often of incalculable magnitude and ... may be in principle uncontrollable.[5]

Our gradual exposition of the nature of the application of causal laws to concrete situations and of the role of statistical investigation will enable us to give a coherent account of these reasons.

Turning now to the more philosophic works which treat of probability and statistics one finds, roughly, either that the discussion occurs within one of problems of induction,[6] or that the central issue has become one of deciding between the 'frequency' and the 'degree of confirmation' interpretations of probability,[7] or the basic concern is to provide adequate axioms for probability theory.[8] None of these orientations exclude, of course, further probing into the questions touched on at the beginning of treatises on probability, but the focusing of interest elsewhere leads to some degree of neglect. A notable exception on the question of randomness is Reichenbach's treatise.[9] Still, from our point of view that treatment requires to be doubly complemented. For it is primarily concerned with the logical formulation of the notion of randomness or irregularity, whereas, firstly, we are more interested in the role of intelligence in coming to that formulation and, secondly, we wish to put the problem of randomness more in

---

[5] *Ibid.*, 4.

[6] For example, G. von Wright, *A Treatise on Induction and Probability*, London, Routledge, 1951.

[7] Cf. E. Nagel, 'Principles of the Theory of Probability,' *Encyclopedia of Unified Science*, Chicago, 1955, vol. I, part 2, 343–422, and the references there.

[8] Cf. P. Suppes, *Introduction to Logic*, Princeton, Van Nostrand, 1957, 274–290, and the references there.

[9] *The Theory of Probability*, Berkeley, University of California Press, 1949, ch. 4: 'Theory of the Order of Probability Sequences.'

the broader context of empirical investigation. It is for this reason that we so formulated and linked the two initial questions which will govern the subsequent discussion of statistics: precisely how do aggregates come to qualify as objects of statistical investigation, precisely how is 'the multiplicity of causes' related to these aggregates and this investigation. The two initial questions will gradually be answered. In the present chapter, however, we will concern ourselves with the raising of some further questions and with a more elementary account of the notion of randomness and of the basic insights of statistics. Chapter eight will deal with more fundamental difficulties relating to the basis of statistical theory.

Two papers on 'Randomness' by G. Spencer Brown and G. B. Keene[10] provide a convenient starting point for our discussion in that the two questions we pose are implicit in their treatment of the first. Brown opens with some bold statements: 'There are no laws of chance. How could there be? Laws predict observations, and it is axiomatic that chance observations are unpredictable by laws. By what then are they predictable? By chance only.'[11] It is to Keene that we owe the clarification and correction of some of the notions involved here, but to Brown the very illuminating way of discussing the problem in terms of 'bamboozling machines': bamboozling in this sense that, taking a penny as a simple chance machine, we find that in a series of ten tosses it can bamboozle, to some extent at least, 1,023 people simultaneously—only one of 1,024 people, who all predict differently the series of heads and tails, will be right.

The significance of our elementary examples and considerations here will only gradually emerge. Eventually, for instance, we will be concerned with answering such questions as, Is the amoeba a bamboozling-machine as far as the physicist is concerned?, Is the world or the process of evolution a bamboozling-machine as far as any human science is concerned?

According to Brown, our concept of randomness is merely an attempt to characterize and distinguish the sort of series which bamboozle the most people. It is only a short step from this to the odd conclusion that 'there are thus no random series,'[12] for someone may turn up a correct prediction. Already we have here a cluster of questions and confusions which require lengthy investigation. The notion, for instance, of chance prediction or 'predictable by chance' will occupy us most especially throughout chapter four. In its regard we may note immediately that Keene succeeds in introducing the important element lacking in Brown's discussion. What makes a prediction chance is not, as Brown would have it, that it is made 'with some lack of confidence' but that it is made on inadequate grounds. 'To predict by chance is simply to predict without justification, where the phrase covers not only the absence of grounds pointing either way, but also

---

[10] Aristotelian Society, Supp. Vol., (xxxi) 1957, Brown 145–50; Keene 151–60.
[11] Ibid., 145.
[12] Ibid., 150.

reliance on insufficient or entirely misplaced grounds.'[13] It would seem, then, that one could permissibly claim a machine to be of limited bamboozling power only if it failed to baffle some grounded prediction. The fact that, in the case of ten tosses, one person in 1,024, all of whom predict differently, will be right after making an ungrounded—chance—prediction is perfectly compatible with the machine being a good bamboozler, with the series being random. Now a prediction cannot be genuinely grounded if there are no grounds for it in the bamboozler itself: one can conclude therefore that 'a randomizing machine is a machine so constructed that there is no reason why any one alternative out of a given set of alternatives, should occur rather than another, at any moment of its operation.'[14] Whatever about the possibility of a randomizing machine—this problem will occupy us especially in chapter three—the factor of most importance about it would seem to be the 'no reason why' which, paradoxically, is required to be a rule of its construction. Again, it is worth noting that this 'no reason why,' this absence of reason, is the common factor in all the usages of the words 'random' and 'randomness.'

We may say, then, that a state of affairs, a situation, is random provided there is a certain absence of intelligibility, of lawfulness, about it. We may note here that the lawfulness involved in our entire discussion has nothing to do with purpose, although this meaning is included in much of the common use of the word 'random.' In such cases random may mean something like 'not sent or guided in a special direction; having no definite aim or purpose; made, done, occurring etc. at haphazard' (O.E.D., B.1). The type of meaning of interest to us, however, also occurs commonly: thus random is 'said of masonry, in which the stones are of irregular sizes and shapes' (O.E.D., B.3). A discussion of this latter case of static randomness will help elucidate our meaning and the significance of randomness for scientific method.

Let us consider, then, the 'situation' constituted by a wall which has been built at random. In the finished wall the stones are situated at random, and these stones are themselves of no particular size or shape. Their size and shape can be said to be any whatever provided specified conditions of intelligibility, regularity, are not fulfilled. How is one to investigate the detailed structure of this wall, the sizes and shapes of its component stones? Were it the regular wall of a standard house there would be little difficulty: one could, for instance, quickly produce a formula for the location of the faces between the regular bricks. But here there is no shortcut. To reach an account of the irregular wall one must move laboriously from stone to stone, taking the details of measurement in each case. Nor is the measurement of one stone of any consequence when one turns to the next— apart from the shape of portions which happen to contact each other. There is

---

[13] Ibid., 154–5.

[14] Ibid., 158.

nothing to warrant the expectation of any relationship between the measurements. Thus one will have at the end of the investigation an aggregate of particular measurements. The aggregate will be coincidental, not held together by any law: it will be a list without a formula. Again, it will be a list without general value. A mason would consider it foolish to carry around with him such a list, as if it were a general prescription for the construction of walls. We note that we have here already examples of coincidental aggregates on different levels: there is the coincidental aggregate of irregular stones; there is the coincidental aggregate of human insights expressed in lists and equations.

The common meaning of randomness can also be specified to some extent by our reaction towards various situations. We arrive at a cocktail party and find on entering the room that by far the majority of the people is gathered at one end. Our spontaneous reaction is to ask Why? : Are the drinks up there, or is the leading bore down here? The spontaneous reaction connotes the expectation that the situation is nonrandom, that there is a reason for the clustering. On the other hand we may find on entering that there are clusters here and there, just as one would expect. For one expects that the distribution of the people would be in some sense random, and if it be a mildly uneven distribution we are not surprised, we seek no explanation for the unevenness.

Already perhaps the specification of randomness shows itself to be somewhat elusive. The problem indeed is that of formulating the peculiar 'there is no reason why' of the quotation from Keene,[15] or the absence of intelligibility which Lonergan refers to when he characterizes a situation as random as 'any whatever provided specified conditions of intelligibility are not fulfilled.'[16] Formulation is normally of reasons, causes, laws or such like: to formulate an absence of reason or law would seem to some extent contradictory. On the level of mathematics this contradictoriness has been explicitly noted. Thus, Feigl remarks of von Mises' specification of randomness that it is 'not mathematically expressible,'[17] and of the same definition A. Church writes: 'while clear as to general intent, it is too inexact to serve satisfactorily as the basis of a mathematical theory.'[18] We may now turn our attention profitably to the problem on that level.

---

[15] Above, p. 14.

[16] *Insight*, 51; CWL 3, 74.

[17] *Erkenntnis* (1) 1930–31, 'Wahrscheinlichkeit und Erfahrung,' 256. More fully: 'So ist der Begriff "Unäbhangigkeit der Auswahl der Teilfolgen von den Merkmalsunterschieden", den für die Definition der Regellosigkeit verwendet wird, mathematisch nicht ausdrückbar. Es handelt sich hier um einen Begriff, der—auch wenn man ihn empirisch zu fassen suchte—nicht in objectiver Weise gerechtfertigt werden kann.' We will deal later with the question of empirical specification.

[18] Alonzo Church, 'On the Concept of a Random Sequence,' *Bull. Amer. Math. Soc.*, (46) 1940, 131.

Let us first consider some elementary illustrations which lead to an appreciation of the origin and nature of statistical understanding.

Consider the differences between the following three simple series:

$$1 \ 2 \ 3 \ 4 \ 5 \ 6 \ 7 \ 8 \ 9 \ 10 \ 11 \ldots$$
$$6 \ 7 \ 5 \ 6 \ 4 \ 7 \ 6 \ 7 \ 5 \ 5 \ 6 \ldots$$
$$5 \ 1 \ 8 \ 4 \ 3 \ 9 \ 2 \ 7 \ 4 \ 6 \ 9 \ldots$$

One understands immediately—and this is usually taken to be the full meaning of the dots at the end—how the first series would be continued. The series is systematic, with a formula for the nth term. The behaviour of the series is lawful and that lawfulness is 'mathematically expressible' in an elementary way. The third series ends also with dots, but the dots added to it have no other significance than as indicating that the series be continued. There is no rule for its continuation in so far as there is no rule relating the first eleven members given. In so far as there is no law relating to it, it may be described as totally random.[19] The terms follow each other in a non-systematic fashion and one does not expect to arrive at a systematic formula governing them or at a generating formula for further members. But when we say that there is no rule for the continuation in so far as there is no rule relating the first eleven members we are not claiming that there is no possible way of arriving at a rule. One could, for instance, use the primitive recursive-function of Gödel to give a type of formula for the eleven values or for any given number of values. For, the β-function is such that for any sequence of natural numbers $k_0, k_1 \ldots k_n$ there exist natural numbers b, c, such that $\beta \, (b, c, i) = k_i$ for $0 \leq i \leq n$.[20] Such a formula is, however, entirely *a posteriori*, it does not suggest further terms of the series, and it is clearly distinguishable from classical generating functions for series. However, there are unlimited ways of taking the eleven numbers as setting a basic pattern in such a manner that further members are indicated. For example, various polynomials in n could be constructed which would give the eleven values indicated and further values besides, these latter values not being uniquely determined. All these rules will differ in an obvious way in complexity from rules such as that governing the first series given. That complexity is shown by the aggregation of particular numerical values, for instance, in the polynomial coefficients or in the β-function structure. One can appreciate that the understanding of such aggregates has not the unity of a single insight, and that lack of unity is in fact extremely significant for foundation questions in mathematics. Here, however, we cannot enter into such foundation questions: our immediate task is to give some introductory clues

---

[19] We are speaking non-technically here. Specialists will recall, for example, K. Popper's attempt to specify clearly *perfect randomness*: Cf. *The Logic of Scientific Discovery*, London, Hutchinson & Co., 1959, 163, footnote *4; also appendices iv and *vi.

[20] Cf. E. Mendelson, *Introduction to Mathematical Logic*, New York, Van Nostrand, 1964, 131.

as to the distinction between the systematic and the non-systematic. While these clues are not adequate to the formulation of the difference between systematic processes and non-systematic processes—since they are concerned with series, not with processes—still, at this stage we might well attempt an incomplete formulation of this distinction. We may define, then, a systematic process as one which can be grasped by a single set of insights so that the investigation of such a process can reach a stage when the data fall into a single perspective and the particulars of any stage in the process can be derived from the particulars of any other stage without the explicit consideration of intervening stages. Now since these are conditions on a type of process, one may define a contrasting process as one in which the conditions do not hold. The understanding, therefore, of the non-systematic process, like the understanding of the non-systematic series, will be multiple, inexpressible in simple unified terms. Further clarification of this definition and distinction may be reserved to chapter three, when the meaning will become more evident through elementary illustration. But the central point is already appreciated if one understands that one's grasp of a function such as a particular Gödel β-function, or even of a simple particular polynomial, has not the unity of a single insight. This is a simple cognitional fact, the appreciation of which requires that one get to grips with the nature of one's own understanding of such expressions. It is quite another matter to move towards an understanding of the ultimate significance of or the basic reason for the different types of insight. That movement will carry us in chapter three into a discussion of primary and secondary components in relations, and in chapter six into the significance of invariance in geometry. But here the elementary fact itself should be apparent. One may point out that a particular Gödel β-function, or a complex polynomial, has at least the unity of an expression. But this unity is the unity of an expression on paper, an empirical unity: what is lacking is unity on the level of understanding. A sign of that lack of understanding might be called the lack of fertility in the formula expressed, and one can consider in contrast the fertility of the insight which gives rise, say, to the generating function for the positive integers, or that which results in the definition of the circle. Here there result formulations which are not tied to particular values and which have virtualities either in the deductive expansion of geometry or in the theory or integers. Again, one may get a similar contrast from the two types of definition of set: definition by function and definition by ostensive methods. It may be noted that in each case the contrast rests in some way on the relation of the systematic to the absence of particularization: to this question we return both in chapter three and in chapter six.

Turning now to the second series we can note immediately that it too is irregular, the terms follow each other in non-systematic fashion, and a general formula for the nth term is not to be expected. Yet the dots at the end of this series indicate more than an arbitrary continuation of the series. The series is less irregular than the third series, and to appreciate that additional meaning is to reach the basic insight of statistical science. Following our policy in this chapter of raising wider issues to be dealt with later, we put this question of the emergence of statistical science into a pseudo-historical context of empirical investigation.

Aristotle did not arrive at the basis of statistical science, although he did appreciate the irregularity of events. That a man be killed by a falling slate in Athens was for Aristotle *per accidens,* and the accidental nature of the event involved the exclusion of a science of men-killed-by-falling-slates. Not that the death of the individual man could not be fully accounted for, by particular trajectories etc.: but the account arrived at would not be of further significance. Upon the death of the next man through a falling-slate accident the account would have to be undertaken again *ab initio.* All this lack of system, formulation, etc. will be discussed later both in the context of physics and in that of biophysics. More immediately we will consider how Aristotle might have been led to bring such accidents to some extent under science by the simple process of counting.

If a second man were killed in a similar way in Athens some months later, there would be little reason to expect a correlation. Still, one might note down that a second such death had in fact occurred. Continuing thus one might find at the end of eleven years that one had a series such as our second series above, each term giving the number of deaths per year. At some stage here, one makes the crucial transition from groping to understanding. The process should remind us of Socrates' sketching in the sand, Aristotle's use of diagram, Aquinas's view on disposing the phantasm, or Hanson's account of patterns of discovery. The counting in itself is only groping. Yet as the number of terms increases the possibility of insight also increases till, through insight, it becomes evident that the series oscillates irregularly round the value 6. The series is irregular yet regular. Here we have what Popper calls 'the paradoxical conclusion from incalculability to calculability,'[21] the fundamental problem of the theory of chance as he calls it, the solution to which 'will be found by analysing the assumptions which allow us to argue from the irregular succession of single occurrences to the regularity of stability of their frequencies.'[22] But where Popper speaks of assumptions we prefer to speak of procedures, where he speaks of conclusions or 'arguing to' we speak of empirically-conditioned insight, where he stresses conjectures we stress gropings in the empirical. Briefly, Popper's treatment of the problem tends deductivistically to exclude the empirical both in its contribution to theory and in its residual character—to this question of residues we will return more than once. Moreover, the procedure of intelligence towards theory is not through a process of conjecture which is beyond investigation but through insight into data which is itself data for an empirical investigation. Popper's discussion of the varieties of

[21] Op. cit., 150.
[22] Ibid., 151.

randomness, however, has the merit of facing certain difficulties relating to verification and empirical reference which, as we shall see,[23] are regularly passed over in the normal textbook treatment.

Let us return to the problem of specifying the irregular yet regular series that are associated with statistics. Von Mises may be said to have initiated the exact discussion of such series,[24] of the problem of distinguishing between various types of series with which might be associated the assumption of the existence of a limit to the sequence of relative actual frequencies of each type occurring in the series. This latter assumption was not involved in the simple illustration given above, but an appreciation of the asymptotic oscillation of relative actual frequencies was implied. Thus by considering the relative frequencies of zeros in groups consisting of the first term only, the first two terms, etc. in the following series of the second type,

$$1 \ 0 \ 0 \ 1 \ 0 \ 1 \ 1 \ 0 \ 1 \ 1 \ 0 \ 0 \ 0 \ 1 \ 0 \ 1 \ 1 \ 0 \ldots,$$

one arrives at this sequence of relative actual frequencies:

$$0, \frac{1}{2}, \frac{2}{3}, \frac{1}{2}, \frac{3}{5}, \frac{1}{2}, \frac{3}{7}, \frac{1}{2}, \frac{4}{9}, \frac{4}{10}, \frac{5}{11}, \frac{1}{2}, \frac{7}{13}, \frac{1}{2}, \frac{8}{15}, \frac{1}{2}, \frac{8}{17}, \frac{1}{2}, \ \cdots$$

The asymptotic oscillation about $\frac{1}{2}$ is evident. It is random, with a randomness governed by the randomness of the initial sequence. The central insight, formulated by von Mises as the assumption of the existence of such a limit as $\frac{1}{2}$ might be here, is that by which one abstracts from the randomness in frequencies to discover regularities that are expressed in the constant proper fractions which are called probabilities. There are problems relating to the assumption of the existence of limits with which we will deal in chapter eight. But it is to be noted that the problem can be avoided on an elementary level by prescribing only that the differences between relative actual frequencies and probabilities be always random: this involves no limit process.

This assumption of the existence of limiting relative frequencies is the first of von Mises' axioms characterizing the series relevant to statistical science. His second basic axiom is the axiom of 'insensitivity to place selection' or of the impossibility of a gambling system. While the two axioms cannot be separated, the second one and the debate surrounding it are more evidently related to the problem of formulating the 'no reason why' which is our concern. The insensitivity referred to belongs to the relative frequencies: they are to remain invariant—with the same values as they have for the principal sequence—for various subsequences. The manner of selecting these subsequences is the crucial issue. Von Mises postulates that any place selection definable by rule belongs to

---

[23] See chapter eight.
[24] R. von Mises, 'Grundlagen der Wahrscheinlickeits rechnung,' *Math. Zs.* (V) 1919, 52–99.

the domain of invariance, i.e. the subsequence thus defined should leave the limits of the relative frequencies unchanged. Now it is objected that this principle of randomness of von Mises is indeterminate, if not contradictory: how could 'a place selection that can be defined by rule' determine a unique class of selections?[25] There are various ways of approaching this difficulty. A first approach is to restrict the class of selections to a well-defined class. This method leads to a defined type of restricted randomness, the randomness of normal sequences. This is the line taken e.g., by Reichenbach, who defines a sequence as normal 'if it is free from aftereffect and if the regular divisions belong to its domain of invariance.'[26] A sequence is free from aftereffect if all the phase probabilities be equal to the corresponding probabilities of the original sequence: phase probabilities have reference to the predecessors (one or many) of the members of the sequence considered e.g., in a sequence of 1's and 0's one may consider the subsequence consisting of members with 101 as predecessors; the probabilities in this new sequence are phase probabilities. By a regular division is meant a division of the major sequence into r subsequences S_{r1}, S_{r2} ... S_{rr} such that to a specified S_{rk} belong all the elements y for which $i = k + (m - 1)r$; $m = 1$, 2, 3 ... ; $k = 1$, or $= 2$, ... or $= r$.

Clearly the normal sequences form a more restricted class than those permitted by von Mises. Indeed it has been shown that it is possible to construct normal sequences by mathematical rule, whereas the von Mises' irregularity defies mathematical formulation. Furthermore—and this is worth noting in reference to our discussion of inverse insight to follow—the set of selections specified by Reichenbach is countable, whereas, as Church points out,[27] it follows from a theorem of Wald that the set of random sequences associated with a fixed probability has the power of the continuum. Another approach to the problem is that of Church himself, in the paper already referred to. Church suggests, as a suitable function to govern selection, an effectively calculable function of positive integers: general recursive therefore according to Turing's result.[28] His basic attitude towards the problem is expressed in the following quotation, which is worth citing in the context and for later reference:

> To a player who would beat the wheel at roulette a system is unusable which corresponds to a mathematical function known to exist but not given by explicit definition; and even the explicit definition is of no use unless it provides a means of calculating the particular values of the function. As a less frivolous example, the scientist concerned with

---

[25] For a more precise related contradictoriness cf. A. Church, op. cit., 131–2.

[26] Op. cit., 144.

[27] Op. cit., 134.

[28] Cf. S. C. Kleene, *Introduction to Metamathematics*, New York, Van Nostrand, 1952, 300.

making predictions or probable predictions of some phenomenon must employ an effectively calculable function: if the law of the phenomenon is not approximate by such a function, prediction is impossible. Thus a *spielsystem* should be represented mathematically, not as a function, or even as a definition of a function, but as an effective algorithm for the calculation of the values of the function.[29]

We will see later, when we discuss classical laws, their independence of particular values and their application to the concrete, that effectively calculable functions are by no means an ideal for the classical scientist. We have already noted how functions of this type can be used to give an apparent systematization of random numerical values. The third series on page 20 occasioned that illustration, when we remarked on the use of Gödel's β-function to give a formula for a sequence of particular values.

Church points out that the existence of random sequences in the sense he suggests is a consequence of theorems of Doob and Wald.[30] But while they are in some ways less objectionable than the normal sequences, the set of random sequences thus defined are even more restricted than the normal sequences of Reichenbach or the equivalent admissible numbers of Copeland.[31]

We may sum up our brief exposition of these efforts to specify randomness with a general comment of von Mises:

> If, instead of restricting ourselves to Bernoulli (i.e. normal) sequences, we consider some differently defined class of sequences, we do not improve the state of affairs. In every case it will be possible to indicate place selections which fall outside the framework of the class of sequences which we have selected. It is not possible to build a theory of probability on the assumption that the limiting values of the relative frequencies should remain unchanged for a certain group of place selections, predetermined once and for all.[32]

Sufficient has been said on this point to reveal the intractability of von Mises' second axiom. What is the source of this difficulty? The central source of difficulty is the contradictoriness of formulating the 'absence of reason' associated with the series in question. The absence of reason here, however, is all

---

[29] Op. cit., 133.

[30] Op. cit., 134.

[31] Ibid.

[32] R. von Mises, *Probability, Statistics and Truth*, London, Allen & Unwin, 1957, 90. Von Mises also criticises the restrictive position in 'Uber Zahlenfolgen, die ein Kollektiv-ähnliches Verhalten zeigen,' *Math. Annal.* (108) 1933, 771–2. Reichenbach (op. cit., 140–50) admits the criticism but considers the normal-sequence specification as sufficient for practical purposes.

the more difficult to handle in that it is an absence where one might expect reasons to be forthcoming—witness the efforts of the mathematicians. This differs from the 'absence of reason,' for example, for the difference between the various occurrences of '6' in the series given. Nobody expects a reason for the difference between two occurrences of the digit '6' beyond the admission that one '6' is here, the other there. There is something empirically residual about this difference which is taken for granted. Under some such category as the empirically residual we may list also what makes scientific generalization possible: when one has explained one hydrogen atom one does not have to find an entirely new explanation for the next. The empirical residue may be described as what is not explained precisely because nobody expects explanation in this case.

All three series given on page 16 exhibit aspects of the empirical residue, aspects which do not come within the realm of explanation or formulation. But the three series, as we have seen, differ further in the way they exclude explanation, formulation, systematization. The three series are in fact listed, so to speak, in the order of increasing perversity. It is the perversity of the second series that is the object of our interest here. It is not totally perverse, there is not a total absence of reason, since it can be brought to some extent under rule. But that bringing to rule has a certain peculiarity, a peculiarity which is paralleled in other parts of mathematics and science. The peculiarity consists in the fact that the rule or law at which one arrives leaves out very clearly but unexpectedly certain aspects of the object of inquiry. Thus the rule which we arrived at with the second series was that it oscillated around the value 6: the rule says nothing about the values of the particular terms although these are obviously part of the object of inquiry. Moreover this omission is a knowing omission: the fluctuations about the number 6 are acknowledged to be lacking in significance. This knowing omission is directly related to the 'perversity' of the object of inquiry and it originates from an insight into that perversity, an insight which grasps that there is nothing to grasp: in that sense, an inverse insight.

One of the clearest illustrations on the level of mathematics of inverse insight occurs in the discussion of the enumerability of the real numbers. Integers and rational numbers are countable but when one adds in the irrational numbers one finds that there is an 'absence of reason' which reveals the rationals to be a negligible proportion of the real numbers in any interval. No procedure of enumeration can succeed here, and this failure to enumerate is crystallized in the diagonal procedure by which one comes to this negative conclusion. The same diagonal procedure is evident in many of the theorems of limitation in Metamathematics.[33] Here again there is a certain expectation—the expectation indeed expressed in the Hilbert programme—which is frustrated, the

---

[33] Cf. Jean Ladrière, *Les Limitations Internes des Formalismes*, Paris, Gauthier-Villars, 1957, 93–140. For the basic antidiagonal notion, cf. 94–5.

appreciation of that frustration being expressed through the inverse insights underlying the concluding stages of theorems like those of Gödel.[34]

On the level of empirical science there is, for example, the inverse insight that gives rise to Newton's first law of motion. The object of investigation is a body in uniform motion in a straight line. One might expect, with Aristotle and the medievals, an explanation of that uniform motion. But Newton's first law crystallizes the inverse insight that there is nothing to explain, and Einstein's position refines that inverse insight further.[35]

The inverse insight fundamentally involved in statistics is not quite as precise as these others. The object of inquiry is aggregates of empirical numerical results. But the denial of intelligibility, the admission of a residue, is not a clear-cut admission like that of the incommensurability of the square root of 2. It is the admission that classical method is frustrated by this object of inquiry. The scientist who is using classical method can never afford to neglect small differences, and in the present case no law can be forthcoming unless small differences are neglected or regarded as lacking in significance. When one is in the classical stance, a small difference in time of orbiting can make the difference of Newtonian and Einsteinian physics. But in statistics a percentage difference is neither here nor there. That the object of inquiry is an aggregate, not a single process or event, is of course of central significance. Thus it is the series of 6's and 7's etc. as series that puzzles, not the individual counts, and the 'absence of reason' admitted by statistics is not an absence of reason in the single events— there is nothing indeterminate about a man being killed by a falling slate—but in the events taken in the aggregate. One moves into statistical method when one adopts the attitude which can acknowledge the difference of the individual terms of the series from 6 to be insignificant. Without that acknowledgement of insignificance, of residue, in the object of inquiry there would be no arriving at the rule, the average. In chapter six we will see how this acknowledged residue is related to a new definition of chance.

Against this background we are in a better position to appreciate the difficulties of those who are seeking to define randomness mathematically. Already we have considered various efforts and their limitations. Popper's 'attempt to find mathematical sequences which approximate to random empirical sequences'[36] is no less limited but more clearly illustrates the flaw in the common approach. Popper develops his view on randomness in the context of the objecttion that criticism of von Mises' axiom of randomness 'seems unanswerable if all possible gambling systems are ruled out,' and so he seeks to define 'chance-like

---

[34] These concluding stages require more explicit expression and methodological reflection than has hitherto appeared.

[35] Cf. B. Lonergan, *Insight*, 23–5; CWL 3, 47–50.

[36] Op. cit., 169.

mathematical sequences.'[37] Popper's efforts, like those of Reichenbach and Church and Copeland, undoubtedly help towards the specification of the object of statistical investigation.[38] Also these efforts are of value to mathematics apart from statistics. But the essential deficiency of the approach rests on a failure to appreciate the significance of von Mises' axiom of randomness, a consequent overstress on 'chance-like mathematical sequences' and a neglect of the genuinely empirical nature of the object of statistical inquiry. One must ultimately acknowledge that the object of this type of investigation is strictly empirical and that it requires to be specified as such, as something-to-be-understood. Thus the problem of the existence of collectives in von Mises' sense can be solved by adverting to the fact that mathematical definition—and so mathematical existence—certainly requires more than von Mises' axiom of randomness, but that nominal definition of its nature lacks clear formulation and yet can be of the existent. There are to be distinguished, then, two types of specification, one on the level of formulation, the other not. Von Mises' axioms of randomness and limiting frequencies fail to measure up to the standards of mathematical formulation, but this failure is not a flaw: it is to be considered as belonging to their role as nominal and heuristic definition. The axioms pinpoint the object of this type of empirical inquiry. The axiom of limiting frequencies is more positive in that it seeks to pinpoint heuristically the key insight in relation to the relevant aggregates. The axiom of randomness is more negative in that it aims at the exclusion of certain series from the inquiry, series which may in fact be governed by the axiom of limiting frequencies but which are also otherwise mathematically definable. It is this latter blanket exclusion of the mathematically definable that has been the bone of contention. But the stress in the axiom is rather on the positive object thus negatively specified. The axiom, one might say, expresses the empirical interest of the inquiry: the series must be the result of a certain type of empirical investigation. In chapter eight we will return to these foundation questions and we will see that there seems to be room for the specification of subtypes of empirical results and for the emergence of corresponding theories of probability.

---

[37] Ibid., 171.

[38] Cf. R. von Mises, *Probability, Statistics and Truth*, 90 ff.

So far our attention has centred on mathematical randomness. The three types of series discussed in that context served as elementary illustrations of what we mean by systematic and non-systematic, the centre series illustrating that curious absence of system which yet displays a certain measure of order.[1] Only in chapter six will we specifically deal with the latter paradox. In the present chapter we make an initial move into the realms of empirical science, making continually more precise the distinction between the systematic and the non-systematic or random.

The second and third of our illustrative series show what is meant by a coincidental or random aggregate on the level of mathematics. Neither of the series can be mastered by a single law, a single formula of classical mathematics. To move into the realm of physics, however, requires that we consider not only non-systematic aggregates of numbers but also non-systematic processes. Our earlier[2] physical example of the random wall did not explicitly involve this latter notion, but an equally elementary example from physics will make clear what is meant.

Instead of an aggregate of stone sizes, then, let us consider the aggregate of locations over a given period of time of, say, ten billiard balls moving on a billiard table. Let us suppose for simplicity that there is no loss of momentum in motion or on impact. In this case one has symmetry with respect to time and it does not matter whether one works backward or forward in time in the deduction of locations. We are of course interested in the deduction of locations, and so we are interested in the application of the simple laws of physics involved—the laws we take to be Newtonian. What we are considering now is not just an aggregate of numbers but of processes. Is the aggregate non-systematic? First of all let us see how physics handles such processes.

Consider the problem of predicting the position of ball A on the table at time t_1 given the positions and momenta of the balls at time t_0. If the time $t_1 - t_0$ is of sufficient length then this problem of prediction requires that one take into account not only the motion of the ball A but the motions of all the balls liable to collide with A. Moreover, there is no *a priori* way of selecting the relevant balls: some certainly may be quickly excluded depending on the shortness of the time $t_1 - t_0$, their velocity and location, but some type of calculation will be required for each. The problem will involve a series of elementary, disconnected calculations. There is nothing like a systematic deduction of the motion of A: one must

---

[1] Above, p. 16.
[2] Above, p. 14.

work tediously from individual collision to collision, determining at each discontinuity the conditions of the subsequent motion. One must, indeed, follow the process out concretely in one's calculations.

The process described has in fact been computerized for 32 balls:

> The idea of Adler and Wainwright who developed this approach at the Lawrence Laboratory at Livermore, was to follow the actual history of the individual spheres as they fly about hitting one another. The imaginary spheres are started off with some arbitrary directions and speeds. The computing machine follows every sphere, keeping track of it, finding where it hits another one, and so on. You can see right away that you cannot do this for very many objects.[3]

Moreover, starting from a position of symmetry is no guarantee of persisting symmetry: 'For example, if you start the system out with all molecules moving at exactly the same speed, after about two collisions per molecule they have all acquired different speeds, distributed in very close agreement with the theoretical distribution curve for an infinite assembly thoroughly random.'[4] Nor, inversely, is there anything to prevent symmetrical positions recurring from time to time during the motion of initially randomly-distributed balls.

Our own viewpoint, of course, was one of predicting, not merely of following the motion. The simple example, however, and the different approaches, serve to illustrate clearly what is meant by a non-systematic process. At this stage we make no claims that such a concrete process is unavoidably non-systematic: we merely show that Newtonian physics can do no better than provide a non-systematic account. We are in fact not establishing here a principle: we aim rather at generating an understanding of the difference between an account of the motion of the billiard balls and, say, the account of the motion of a planet. Moreover, this physical illustration of the non-systematic is more easily appreciated than the mathematical case considered in chapter two. There we had a list of numbers which, fundamentally, could not be grasped in a unified fashion: the unity which they might assume either through use of something like Gödel's β-function or more simply as the roots of a single polynomial is not the unity of a single insight. In the physical case we have more evidently the exclusion of that type of unified understanding: we have a sequence of lists of definite integrals of equations of motions and an aggregate of boundary conditions, collision coordinates, etc. Most evidently we have geometrical considerations and partic- ular coordinates, and, as chapter six will reveal, there is a clear relation between the systematic, the laws of physics and the invariants of geometry on the one

---

[3] Edward Purcell, 'Parts and Wholes,' in *Parts and Wholes*, edited by Daniel Lerner, London, Macmillan, 1963, 19.

[4] Ibid.

hand and on the other hand the non-systematic, the particularization of laws of physics and particular coordinates.

The contrast we are making is between systematic and non-systematic processes, between systematic and non-systematic understanding. To appreciate that contrast sufficiently at this stage requires, as we pointed out in the case of the mathematical illustrations of chapter two, that one advert to one's experience of the type of understanding, calculation and prediction that go with either process. We may note in passing that the processes as we consider them are ideal: this is evident both from the ideality of the conditions of motion in our considerations and from the general condition of isolation implicitly imposed: to this feature of our understanding of concrete process we will return presently. Let us compare then the procedure of prediction of the position at any time of a body moving in an elliptic orbit round a centre of force with the problem of prediction of the position at any time of one of the billiard balls in our illustration. Instead of the general formula from which one derives, as it were at a stroke, the position at any time of the orbiting body, one has in the case of the billiard ball an aggregate of equations and conditions from which one can derive only with tedium later positions of the particular billiard ball.

Note that the difference between the two processes does not rest primarily on a difference of types of initial conditions but only on these in so far as they ground interference. Initial conditions indeed can themselves be an unsystematic aggregate yet the process can be systematic: thus the positions and velocities of the planets are not systematically related, yet the whole process can be treated systematically: the single set of insights involved in determining the set of initial conditions together with the relevant laws are sufficient to give knowledge of the whole motion. On the other hand there might well be, as our illustration shows, a certain symmetry of initial conditions without the process being systematic, without the set of conditions together with the laws involved being adequate for a deduction of the whole process. The distinction between the two derivations would seem from this, perhaps, to be one only of degree of difficulty, or to rest only on the difference between non-collision and collision. But closer analysis will reveal that the distinction between systematic and non-systematic processes is more basic, as indeed is the difference between processes involving or not involving collisions. One can note immediately, for example, that the question of collisions will bring us close to Aristotelian discussion of the coincidental or the *per accidens,* and the ultimate possibility of collisions will be associated with the basic category of the empirical residue. These are questions with which we will deal later.

Let us return to the problem of prediction in our simple example of a non-systematic process. To present in detail the concrete derivation of a particular process would be tedious. Nonetheless, because the elementary contrast can be missed, it is worth noting for oneself how one must proceed by insight into the concrete process when one is trying to derive, say, the location of ball A after 4 seconds on the table, given that at time t_0 ball A is at position $(1,1)$—with some convention of axes—with velocity $(1,1)$, ball B is at position $(3,5)$ with velocity $(0,-\frac{1}{2})$, and for simplicity the other balls are assumed to be too far removed to interfere in the interval of 4 seconds. If the balls have diameter of one unit, then one derives a collision position at time $t = 2$, at which time ball A is at point $(3,3)$, and ball B is at point $(3,4)$. One must proceed then to calculate the changes in velocities etc., and with these conditions determined discuss the further motion.

Already in chapter two we have characterized the contrast between systematic processes and non-systematic processes more generally. The meaning of the characterization should now be clearer, most especially the key factor of how the account of a non-systematic process is, one might say, continually bedeviled by particularity. One discovers the location at some given time t of some definite billiard ball, not by putting a set of values into a, general formula, but by taking account of a sequence of particular collisions and motions. Again one may note that the account of a non-systematic process will fit only processes identical with itself in initial conditions etc. A small change in these initial conditions would demand, not some minor adjustment of the account, but perhaps an entirely new account. Furthermore, no account of this type would be considered to be of general scientific value, no more than the list of stone sizes discussed in chapter two. Thus, having laboriously worked out the total motion of the system for a definite length of time with definite initial conditions, one is in no better position to deal with any other like system, unless it have identical conditions. If it has not, then one must start from the beginning. The fact that such an account of a particular system is not of general value will take on a new significance when we come to deal with the problem of systematic understanding of biological phenomena in chapter nine. There we will show that the total biochemical and biophysical account of one particular amoeba is multiply-non-systematic and entirely particular, while the biological understanding of the amoeba is both systematic and general. In the present context, however, our interest is in the general knowledge of non-systematic processes which can be gained through statistics. Statistical science gives a general account, not of the process itself, but of boundary values and such like associated with the process. The reason for this and its significance will gradually emerge.

Finally, we must note that being non-systematic and being indeterminate are not equivalent. There is no question of the process just considered being indeterminate: the motion in the simple illustration of the billiard balls is determinate and might be determined down to its last detail. In the context of that distinction we may move to another aspect of the central problem of this chapter as it relates to Laplace's view on probability theory and determinism.

We may put the problem in the words of Jevons: 'Chance then exists not in nature, and cannot coexist with knowledge; it is merely an expression, as Laplace remarked, for our ignorance of the causes in action ... Probability belongs wholly to the mind.'[5] On these terms one might even concede to statistics an interim usefulness—a considerable usefulness, as will appear in chapter five—and still claim that statistics are only a cloak of ignorance. That claim looks to a future date—'infinitely remote' as Laplace grants in the following quotation—when world process will be understood in classical terms, even perhaps 'in one single formula.' As Laplace is the central representative of the view we wish to reject in detail, we had best quote his own clear detailed statement of the claim:

> We ought to regard the present state of the universe as the effect of its antecedent state and as the cause of the state that is to follow. An intelligence knowing all the forces acting in nature at a given instant, as well as the momentary positions of all things in the universe, would be able to comprehend in one single formula the motions of the largest bodies as well as of the lightest atoms in the world, provided that its intellect were sufficiently powerful to subject all data to analysis; to it nothing would be uncertain, the future as well as the past would be present to its eyes. The perfection that the human mind has been able to give to astronomy affords a feeble outline of such an intelligence. Discoveries in mechanics and geometry, coupled with those in universal gravitation, have brought the mind within reach of the system of the world. All the mind's efforts in the search for truth tend to approximate to the intelligence we have just imagined, although it will forever remain infinitely remote from such an intelligence.[6]

We may immediately grant that Laplace's position would be at least partially acceptable if in fact the universe were as systematic as an ideal realization of the solar system. In that case knowledge of a few strategic conditions might well make possible accurate predictions of future states. But even then his position would only be partially acceptable. For one thing, a distinction between knowledge and prediction is relevant: 'the future as well as the past would be open to its eyes' is simply not true. The predictor's relation to the future particular is indeed a complex problem which occupies Aquinas regularly because of its relation to the problem of divine knowledge.

> The astronomer can predict all the eclipses of coming centuries; but his science as such will not give him knowledge of any particular eclipse as particular, *"sicut rusticus cognoscit"*; for in so far as the

[5] *The Principles of Science*, New York, Dover, 1958, 198.
[6] *Théorie analytique des probabilités*, Paris, 1820, préface.

astronomer knows future eclipses as particular, it is only by relating his calculations to a sensibly given here and now.[7]

We may remark in passing that the problem of divine knowledge is not our concern here: at all events, since divine knowledge is neither systematic nor predictive, for it difficulties of systematization of a non-systematic universe do not arise.[8]

In touching thus on the problem of knowledge or prediction of the particular we are in fact very near the heart of the matter. We recall here Brown's initial statement quoted earlier, 'Laws predict observations.'[9] This simple statement cloaks a variety of important and relevant problems of the relation of laws to particular events which we must investigate prior to dealing adequately with Laplace's notion of 'one single formula' or of the systematization of a deduction of world process. To bring the problems out into the open we had best take a closer look at actual scientific practice in the process of prediction.

The application of causal or classical laws to concrete situations is a complex process involving a set of insights the methodological significance of which is consistently ignored. First of all, classical laws are general: one has only to page through any standard textbook in physics to appreciate this. The laws may include particular numerical values but not particular boundary conditions: these are brought in only through the application of the laws to concrete situations.[10] Indeed, modern presentations of physics tend to use increasingly the technique of implicit definition which stresses this remoteness from particularity and the concrete. Thus, for example, $\mathbf{E}$ and $\mathbf{H}$ are defined implicitly by the electro-magnetic equations, and fundamental experimentation may be taken to provide corresponding descriptive concepts.[11] Again, the set of thermodynamic principles and laws define implicitly the terms they contain, and in a systematic account of thermodynamics one might well 'develop a purely mathematical theory in which the terms *state, +, →, entropy, component of content,* and so on, appear as mathematical objects without any physical connotation. One can then attach to this exposition

---

[7] B. Lonergan, *Verbum: Word and Idea in Aquinas*, Notre Dame, 1967, 40; CWL 2, 53. ['as the simple peasant knows']

[8] The classic passage in St. Thomas on divine knowledge is *Peri Hermenias*, lectio 17; cf. also B. Lonergan, *Insight* ch. 19, esp. 649–51; CWL 3, 672–674, D. Bohm, *Causality and Chance in Modern Physics*, New York, Harper, 1957, 159.

[9] Chapter 2, 13.

[10] N. R. Campbell's book, *Physics: The Elements*, Cambridge University Press, 1920, raises many of the questions which will concern us here, especially his discussion of measurement and numerical laws in Part II.

[11] Cf. Lindsay and Margenau, *Foundations of Physics*, New York, Dover, 1957, 303, 306.

a "text" explaining the physical meaning which is to be assigned to the various terms employed.'[12] The mathematically-meaningful differential equation

$$\frac{\delta^2 V}{\delta x^2} + \frac{\delta^2 V}{\delta y^2} = \frac{1}{c^2} \frac{\delta^2 V}{\delta z^2}$$

provides a more particular illustration. It takes a specific empirical meaning when one is told that it is applicable to the harmonic vibrations of a uniform plane membrane, say a circular drum top, V being the displacement at time z, c being a constant depending on the density and tension of the membrane. The equation thus embodies certain laws of physics, but it is clearly general, as are the equations implicit in the previous two illustrations.[13]

The passage from a knowledge of general laws to particular cases calls for an intelligent selection of the laws suspected to be relevant to the particular cases, their suitable combination and their particularization through the replacement of arbitrary parameters etc. by numerical values pertaining to the concrete situation. The scientist will easily recognize this as standard procedure although neither he nor the philosopher may appreciate its significance especially in the present question of the objectivity of statistical knowledge. The procedure may be simply illustrated by the problem of making predictions in the case of plane oscillations of a simple pendulum. Here laws are selected which are considered relevant: certainly the law of gravity in some form, possibly laws of air resistance or axle friction, etc. The laws in this case can be combined in a single differential equation[14] and the integration yields a general solution which must be particularized by the scientific observation of the particular pendulum and the initial conditions of its motion. Only then do predictions of the future states of the system become possible. These predictions, moreover, are conditioned predictions, the conditioning being normally acknowledged, at least implicitly, by a blanket proviso, 'other things being equal.' The nub of our argument will consist in teasing out the full significance of this blanket proviso. But before coming to that nub, various other difficulties have to be eliminated.

To come closer both to these difficulties and to the region of statistics we will consider the same problem of the application of classical laws to the case of the tossing of a penny. Again we have the task of selecting the relevant laws, laws

---

[12] F. Giles, *Mathematical Foundations of Thermodynamics*, New York, Macmillan, 1964, 5; his first chapter is a good discussion of this problem, but whether his system is all that he claims it to be is questionable.

[13] The full significance of these remarks will appear when they are taken in conjunction with the treatment of abstraction to follow and with the discussion of nominal and implicit definition in chapter 8.

[14] Differentiation is an abstractive procedure. Cf. Lindsay and Margenau, *Foundations of Physics*, 29–55, where they discuss 'the method of elementary abstraction' as yielding differential equations.

such as those listed in Arley and Buch's discussion of the throwing of a die.[15] The combined laws yield a set of differential equations governing the motion; integration and the determination of constants and initial conditions put us in a position to predict the final state of the penny—all the more easily in that in practice there are only two relevant final states.

First of all let us say that, while conceding Arley and Buch's point[16] regarding the difficulties of determining initial conditions, we will assume for the sake of argument that initial conditions for any throw can in fact be exactly specified. The impossibility of such a specification would in any case strengthen rather than weaken our case. In making this assumption we are taking the same line as Reichenbach; we will see immediately where the weakness lies in his conclusion regarding exactness of prediction:

> We say, for instance, that observations of the initial velocity of the spinning ball do not enable us to make a deviating selection of the results and that thus a selection based on such observations belongs to the domain of invariance of the sequence. But this is true only in view of the limited abilities of human observers, as far as both observation and mathematical computation are concerned. With precise observation of the initial velocity of the spinning ball, it should be possible to foretell with any degree of exactness where it will come to rest. It is only technical ability that prevents us from equaling Laplace's superman.[17]

Given that all the relevant laws are known to us, we can agree with Reichenbach that a sufficiency of technical ability would enable us—whether we play 'pitch and toss' or roulette—to observe initial conditions, make rapid calculations and place our bet on a result foreseen, prior to the landing of the penny or the call of the croupière, 'rien ne va plus.' But the identity of the result with the result foreseen is conditioned, with a conditioning which remains operative right up to the occurrence of the result. That conditioning has already been expressed by the general phrase 'other things being equal.' In terms familiar to the scientist, it is the assumption of adequate isolation, and in practice the fact that other things are not 'equal' is allowed for with a required degree of precision, the allowance varying from complete rejection of experiments in which major disturbances were effective, to the inclusion of corrections for temperature variations etc. Hempel puts the point well:

> For certain purposes in astronomy the disturbing influence of celestial objects other than those explicitly considered may be neglected as

---

[15] Cf. above, pp. 11–12.

[16] Ibid.

[17] H. Reichenbach, *The Theory of Probability*, 150.

insignificant, and the system under consideration may then be treated as "Isolated"; but this should not lead us to overlook the fact that even those laws and theories of the physical sciences which provide exemplars of the deductively-nomological prediction do not enable us to forecast certain future events strictly on the basis of information about the present: the predictive argument also requires certain premises about the future—e.g. absence of disturbing influences, such as a collision of Mars with an unexpected comet—and the temporal scope of these boundary conditions must extend up to the very time at which the predicted event is to occur. The assertion therefore that laws and theories of deterministic form enable us to predict certain aspects of the future from information about the present has to be taken with a considerable grain of salt.[18]

In the following chapter we will expose the flaws in Hempel's own treatment of statistical science. Here, however, we are in agreement, and our immediate concern is with specifying exactly how considerable the 'grain of salt' alluded to is.

There is one further point to note regarding the penny-tossing illustration which helps to focus the main difficulty. Even if the fulfilment of the condition 'other things being equal' is granted for a sequence of penny tosses, we are no nearer the solution of the problem of systematization. Thus, the granted condition enables the gambler in the present hypothesis to predict exactly the final states, given that the initial conditions are observed in each case, or better, prearranged and listed beforehand. In the latter case the gambler could have a function of the type specified by Church[19] for initial or final states, but even then a sweeping, systematic, prediction, of the sequence of results would be out of the question. The deduction of the final states, however rapidly carried out by Reichenbach's superman, would be piecemeal, non-systematic. Moreover, the possibility of a systematic mathematical transformation which, operating, say, on something like a β-function of the initial states of the penny, would yield directly a β-function of the conditions of the final states, is excluded by the non-systematic nature of the two aggregates involved in the initial and final β-functions, and of the relations between the two sets of numbers specifying each particular toss.

We must conclude then that a systematic account of the aggregate of processes associated with a sequence of coin tosses is out of the question under the conditions mentioned. If the same can be said with the blanket condition removed, then the case for the objectivity of statistical science can be solidly founded. For, that case rests on the objectivity of the non-systematic. If there are

---

[18] C. G. Hempel, 'Deductive-Nomological vs. Statistical Explanation,' *Minnesota Studies in the Philosophy of Science*, Minneapolis, 1962, edited by H. Feigl and G. Maxwell, 116–7.

[19] Cf. chapter 2.

processes which radically elude systematization, then there is scope for statistical understanding even when all classical laws are known. The absence of system is not a complete exclusion of general knowledge. As our paradigm case of the second series shows, in certain non-systematic situations statistical science can, so to speak, slip through the gap of unlawfulness to supply such a general knowledge.

But perhaps the removal of the condition 'other things being equal' and the inclusion not only of conditions such as the initial position and velocities of the penny but of all relevant factors and influences would open the way to systematization, to a complete coherent and systematic account of the processes involved in an evening's gambling?

Certainly for Aristotle there was no question of achieving such an account, and while he was unaware of the possibility of the general knowledge available through statistics, his reason for excluding systematic demonstration is, as we shall see, basic and permanently valid: 'The more demonstration is particular, the more it sinks into an indeterminate manifold, while universal demonstration tends to the simple and determinate universal demonstration is intelligible, while particular demonstration verges on sense-perception.'[20] In contrast with Aristotle's position there are the views of, for example, Poincaré or Kneale—the latter's view will be considered in chapter six in a complementary context—which would incline one to hope for the ultimate achievement of the account in question. This is true of Poincaré's view in so far as his cataloguing of facts might become systematic:

> Let us compare science to a library that ought to grow continuously ...
> It is experimental physics that is entrusted with the purchases. It alone,
> then, can enrich the library. As for mathematical physics, its task will
> be to make out the catalogue. If the catalogue is well made, the library
> will not be any richer, but the reader will be helped to its riches.[21]

The catalogue is of scientific facts but 'the scientific fact is only the crude fact translated into a convenient language'[22] and the ultimate goal would be a complete catalogue, for, in Poincaré's position 'there remains ... not the smallest place for anything whatever that could be called contingence.'[23] The categories of systematic and non-systematic are not, of course, used by Poincaré, and our citations are rather indications of a general tendency than a precise interpretation of Poincaré's position. But the basic assumption of the general tendency would

---

[20] *Posterior Analytics*, Bk. 1, cp. 24.

[21] *Science and Hypothesis*, 130. Page references in this and the following two footnotes are to *The Foundations of Science*, New York, The Science Press, 1946, second edition.

[22] *The Value of Science*, New York, Dover, 1958, 330.

[23] Ibid., 343.

seem to be clearly identifiable, and it is an assumption which is in opposition both to the Aristotelian view and to scientific practice.

The crucial issue is the question of abstraction, and the opposition mentioned is between a view of abstraction as adding to data or enriching data and the view which overstresses the negative aspect of abstraction. On this latter view, abstraction yields some type of impoverished replica of concrete data. This view fits neatly into the 'library' view of science. If one assumes that every aspect of sensible data has its impoverished replica then the aim of science is to catalogue these in some way. If one succeeds, then from the totality of impoverished replicas one may expect to be able to reconstruct the totality of sensible data. Furthermore, the claim that statistical laws are merely a cloak of ignorance amounts, on this view of abstraction, to the claim that all the replicas are systematically related: for, as we have seen, where there is absence of system, there is scope for statistical science. Again, on this view, the systematic relations between the impoverished replicas must also hold in the concrete, and so the concrete is not just determinate—as we would in fact hold—but strictly systematic.

The opposing view of abstraction on which our case ultimately rests is the much misrepresented view of Aristotle.[24] This view is that abstraction, far from being an impoverishment of data is rather an enrichment of data: it is not a subtractive process but an additive process.[25] Moreover, this abstraction is not an unconscious quasi-mechanical process, the *simplex apprehensio* of the later scholastics, the *conceptio* of Scotus, yielding concepts which require later analysis that they be understood.[26] It is a conscious process which questioningly anticipates an intelligibility to be added to sensible presentations, which searches

---

[24] The misrepresentation of the Aristotelian position runs right through Thomism, a school which would claim Aristotelian parentage. On the later scholastics cf. Fr Hoenan, 'De Origine Primorum Principiorum Scientiae,' *Gregorianum* (XIV) 1933, 153–184; (XIX), 498–514; (XX) 1939, 19–54, 321–50: cf. also B. Lonergan, 'St Thomas Theory of Operation,' *Theol. Stud.* (3) 1942, 374–402. For the modem scholastics and a correct representation of the position of Aristotle and Aquinas cf. B. Lonergan, *Verbum: Word and Idea in Aquinas*; CWL 2. Cf. also the references in footnote 31 of chapter six below, 89.

[25] Various non-Thomist authors, to which we will refer in other parts of this work, are beginning to appreciate the significance of this. In this context one might consider the illustrations given by N. Hanson in *Patterns of Discovery*, chapter 1; also 36ff, on Galileo's and Descartes' transcending of conceptual limitations, and 72ff on Kepler's discoveries: '. . . the elements of an inquiry coagulate into an intelligible pattern. The affinities between seeing the hidden man in a cluster of dots and seeing the Martian ellipse in a cluster of data are profound.' Ibid, 86.

[26] Representative of such an approach is G. Hempel's 'Fundamentals of Concept Formation in Empirical Science,' *International Encyclopedia of Unified Science*, vol. 2, no. 7, Chicago, 1952. From the present point of view he has passed over *the* fundamental.

out the significant, the essential, what Aristotle would call the form,[27] and which only terminally reaches the theory, the definition, the formulation, the *verbum incomplexum* of Aquinas.[28] We will return to a discussion of Aristotle's view in another context in chapter six. But we note here that while it is a view of abstraction which runs counter to much contemporary discussion of human understanding and scientific knowledge, it rests solidly on scientific practice and nonscientific experience. Here we recall Einstein's maxim about considering not what the scientists say about their science, but what they do. The acceptability of either view of abstraction can be left to the courts of scientific practice.

The previous paragraph began by pointing out that our case rested on Aristotle's view of abstraction and concluded with an appeal to scientific practice, but the paragraph indicated that we thus have not two courts of appeal but one. Obviously here we are touching on fundamental issues in philosophy and in the interpretation of Aristotle. The indications given just now will in fact be developed in the concluding chapter into a view of philosophy which provides the ultimate vindication of our thesis. As was pointed out at the end of chapter one, in so far as that view is not shared, our arguments will be considered inconclusive. So, for example, the methodological nature of the Aristotelian notions to which we refer will be missed, and so their relation to the practical procedures of science. Then our conclusion regarding statistical science as providing general knowledge of non-systematic or random processes will be considered as merely a conclusion

---

[27] There are complications here which we pass over, e.g., the nonidentity of form and definition. Still, we may make the general point that the scientists' goal is form, theory—verified, of course: he seeks the form of nuclear forces, the form of development, the structure, form, of the evolutionary process, the structures of societies, etc. There has been a tendency, not restricted to scholastic metaphysics, to turn 'form' into a myth: form then becomes something which metaphysicians alone intuit. To this mythmaking corresponds, to some extent, an overconcern with efficient causality. But the practising scientist continues to push towards an understanding of form, whatever he may call it. 'The tracts, treatises and texts of the last three hundred years of physics rarely contain the word *cause*, much less *causal chain*. In their prefaces and their *obiter dicta* physicists may get expansive; nonetheless the concept is used infrequently in the actual practice of physics, and this fact is important.' (N. Hanson, *Patterns of Discovery*, 51–2); 'the primary reason for referring to the cause of x is to explain x.' (Ibid., 54) The scientist, one may say, takes for granted relations of efficient dependence: what he is interested in is their form.

[28] Relevant texts of Aquinas are collected in B. Lonergan, *De Deo Trino*, Vol. II, Rome, 1964, 273–290; CWL 12, 569–587. Aquinas' position is summed up in the text, 'Quicumque enim intelligit, ex hoc ipso quod intelligit, procedit aliquid intra ipsum, quod est conceptio rei intellectae, ex vi intellectiva proveniens, et ex eius notitia procedens' (*Sum. Theol.* I, q. 27, a.1). ['Whenever we understand, by the mere fact that we do understand, something proceeds within us, which is the conception of the thing understood, issuing from our intellectual power and proceeding from its knowledge.']

about scientific practice which lacks philosophical significance. While it would be good to eliminate immediately the appearance of inconclusiveness, we prefer to postpone that task till we have the background of the larger and related context of chapter nine. It is in chapter twelve that the argument falls into adequate perspective. We proceed immediately, then, to enlarge our cumulative argument by drawing on instances of scientific practice which are identically instances of Aristotelian abstraction. But the argument genuinely accumulates and the identity is alone appreciated within the adequate perspective, which is one of scientific self-attention on the part of the reader.

We will take a simple illustration as paradigm for the procedure in science: the discovery of the law $PV = C$. Already we have remarked on the generality of all such laws in science, a factor closely related to the present issue, but also to the question of the empirical residue discussed earlier.[29] The need to admit an empirical residue, indeed, reveals the first flaw in the impoverished-replica-library view of science: for it is an aspect of data that in fact yields up no impoverished replica.[30] It is this which underlies Aristotle's view when he remarks that 'the more demonstration is particular the more it sinks into an indeterminate manifold.' Inversely, one moves from the indeterminate manifold by abstraction, but that abstraction has not only the negative aspect of admitting an empirical residue, but the more important aspect of enrichment. How that proceeds can be concretely illustrated in our simple case of the ideal gas law. That law, far from being a relation between impoverished replicas, can be seen to result from a correlation of correlations of correlations of data. The triple 'correlation' is deliberate. The key to the movement towards the law lies in measurement, and the initial correlation results from the selection and use of a basis of measurement. There is then the process of correlation involved in providing suitable measuring apparatus. There is next the sequence of correlations of measuring apparatus with measured which yields in our case a list of volumes or a list of pressures. The lists are at this stage materially associated. Finally, the central insight giving rise to the law, the correlation, is that which associates the two lists—abstracting concomitantly from errors and random differences—to arrive at the relation of inverse proportion. The scientist will have no difficulty in recognizing the description of the procedure, which is typical of the emergence of any scientific law. That scientific law is abstract, general, making no mention of particular cases or particular values.

The basic distinction indeed is between abstract system and particular cases. Both can be understood, but they are understood in different ways. To know the particular case requires sensible presentations: this is illustrated clearly by the procedure of determining initial conditions for an oscillating pendulum or for the

---

[29] Cf. above, p. 22.

[30] We will return to this, in discussing Kneale's position, in chapter six.

tossing of a penny. But the abstract system is neither sensible nor imaginable—no more than one can imagine the points and lines defined by Hilbert. No definite numerical values are involved except in so far as these enter into the definitional correlations. The scientific law gives, therefore, only what may be called primary correlations: for example, the correlations implicit in the electromagnetic equations for **E** and **H,** or the correlations implicit in a gravitational relation field for position and momenta of masses.[31] To pass from these primary correlations to the secondary determinations of, say, how many masses with what momenta are in what positions, one must appeal to observation and measurement of particular cases. In practice in physics this procedure is associated with integration, for just as differentiation is an abstractive process[32] so integration normally corresponds to a return to the concrete. Integrals can of course be left with the generality of arbitrary constants: then one has what might be called indefinite explanatory models e.g., the general elliptic orbit equation.

The secondary determinations, at all events, do not come within the direct scope of the classical laws. To use Hempel's analysis,[33] they belong to the minor premise of the general syllogism of scientific inquiry. These secondary determinations—particular velocities, particular energies, particular heights etc.—are therefore not included in the classical systematization. When they are included, as they must be for prediction or for model-construction, they are in general listed, extensionally defined.[34] The case for real randomness rests on the claim that these secondary determinations cannot in general be intentionally defined or systematized. The case for statistical science rests on the fact that they can nonetheless be brought to some extent under rule.

The foregoing analysis shows clearly the unacceptability of the view, implicitly advocated by Poincaré and by the adherents of an impoverished replica position on abstraction, that argues for an isomorphism between classical laws and concrete data. Firstly, because abstraction in the case of classical laws is negative in prescinding from empirically residual aspects of data, one cannot argue from the systematic nature of such laws to the systematic nature of the concrete data: for, the effort to account in detail for the concrete involves the addition of conditions which may not be systematic. Secondly, because abstraction goes beyond sensible data to clearly and, usually, implicitly defined

---

[31] This is amply illustrated by standard general treatises, for example, Whittaker's *Analytic Dynamics.* On the question of primary and secondary components in relations and their significance for the distinction both between classical and statistical inquiry and between the systematic and the non-systematic cf. B. Lonergan, *Insight*, 490–7; CWL 3, 514–520; *De Deo Trino*, Vol. II, Rome, 1964, 291–315; CWL 12, 687–711.

[32] Cf. footnote 14 above, p. 31.

[33] Op. cit., footnote 18 above, p. 33; and Hempel, op. cit. 100. We will return to this analysis in chapter 4.

[34] Cf. H. Reichenbach, *The Theory of Probability*, 340–2.

terms and relations, the correspondence of classical laws and concrete situations is nothing like a one-one affair. The classical laws arrived at by this negative and positive abstraction cannot be put into any simple isomorphism with particular situations. Arriving at a law is a complex process of correlation and abstraction from random differences, and the inverse process of verification is an equally complex process. What is ultimately verified is, not a particular set of values, but the general abstract formulation. Moreover, that verification is had, not by some coincidence of 'scientific fact' and 'crude fact,' but by the convergence of experimental results on the abstract law.[35]

Complete knowledge of all classical laws would give knowledge only of such abstract laws. One may object, however, that in so far as science moves towards an integration of all such laws it moves towards an integrated, systematic account of the universe. It has indeed often been pointed out that science aims not merely at laws and combinations of laws, but at ever more comprehensive theories.[36] Just as one may pull together into a single equation the laws of gravity and friction and air resistance in the case of the pendulum, just as one may bring into a single set of equations gravitational and electromagnetic theory,[37] so one may hope to bring together in some masterly synthesis, 'one single formula,'[38] an explanation of all phenomena. Instead of one unique formula one may consider one unique model to be the goal of scientific inquiry: then the stress is on the role of models in science and the hope of continually inventing more all-inclusive models.[39] The two goals differ as abstract synthesis and concrete synthesis. The question of abstract synthesis need not delay us: one has only to inspect any of the proposed sets of differential equations for a four-dimensional curved manifold to appreciate how removed they are from concrete details of number and distribution.[40] Again it may be true, as Heisenberg says, that 'the totality of Schrödinger's differential equations corresponds to the totality of all possible

---

[35] We will return to the question of verification in chapters 7 and 8.

[36] Cf. for example Nagel, *The Structure of Science*, London, Routledge, 1961, ch. 5.

[37] Cf. Schrödinger, *Space Time Structure*, Cambridge University Press, 1950, ch. 12.

[38] The phrase, and the hope, occur not only in determinist contexts—cf. the quotation from Laplace, above, p. 29—but also in indeterminist circles.

[39] Both the types of models and their uses are legion. The role of models in science is discussed, e.g., in M. B. Hesse's *Models and Analogies in Science*, London, Sheed & Ward, 1963; also in R. Harré's *Theories and Things*, esp. in section 3. There is a parallel between the rejection in this latter work of a positivist principle 'All bridge-statements are identities, not analogies,' 24, and our rejection above of 'impoverishing abstraction.' Cf. Also footnote 68 of chapter 10 below, 186.

[40] Even Eddington, whose *Fundamental Theory*, Cambridge University Press, 1953, concludes with the total number of protons and electrons in the universe (283), makes no attempt to give details of their distribution.

states of atoms and chemical compounds,'[41] but this is still a far cry from a detailed account of the actual states. Only an unsystematizable multitude of observations could begin to bridge the gap.

At first sight a concrete synthesis through an adequate model seems very much nearer to the desired integration. Such a model would seem to include not only the abstract synthesis but also the concrete particular details: the paradigm case would be a micromodel of the solar system. From our point of view, the most obvious flaw in the 'adequate model' argument is that in this case the particular conditions are not included systematically: they are included concretely, built into the model. One can build a model of the solar system without having a systematic formula, for example, for the distances of the nine planets from the sun: all one needs is a list of this coincidental aggregate of distances.

Continually we return to this central point: the existence of coincidental aggregates of values and processes. The discussion of the abstract nature of classical laws began from the suggestion [42] that one might perhaps reach a systematization of such aggregates if one removed the condition 'other things being equal' from the normal prediction procedure, and took everything into account. That discussion of classical laws has shown that one can take all classical laws into account and still leave out of the account the secondary determinations. Moreover, a systematization of all or some classical laws fares no better, while a concrete synthesis or model does not require any systematization of particular values. The full force of the argument for the non-systematizability of secondary determinations appears, however, when one follows out the suggestion of removing the condition 'other things being equal.' For then one reveals that this general condition cloaks a divergent expansion of non-systematic conditions. The citation from Hempel above[43] hints at this, but the intervening discussion puts us in a position to understand it accurately. As that discussion shows, the blanket condition does not connote the concentration on certain particular 'impoverished replicas' to the exclusion of the consideration of others. It connotes rather that classical laws—such as that for motion under gravity in the case of coin tossing— hold in the concrete only in so far as a range of conditions are fulfilled. The actual motion occurs in the 'indeterminate manifold' of which Aristotle speaks, and that indeterminate manifold is the possibility of a multiplicity of intersections and conjuncttions of classically-determined motions.[44] Actual consideration of the secondary determinations of classical laws brings one into this manifold. Hence, to explain adequately any particular secondary determination, for example, why

---

[41] W. Heisenberg, *Philosophical Problems of Nuclear Science*, London, Faber and Faber, 1952, 101.

[42] Above, p. 34.

[43] Above, p. 33.

[44] Cf. footnote 24 of chapter 6 below, p. 88, and the text there.

the Earth is at a particular space-time interval from the sun, involves one in the concrete deduction of the possibly influential processes.

At no stage did we deny the possibility of such a concrete deduction. Our central thesis is rather that such a concrete deduction is non-systematic. To put the matter another way and in terms of our basic illustration, we do not deny the possibility of working backwards concretely through the determinate antecedents of the sequence of results of a night's penny tossing: we merely deny that the proceedings of the night can be systematized.

Let us consider the problem of tracing back the antecedents of one particular result of penny tossing. As we conceded earlier we could in principle deduce this result from the initial conditions of the toss and the classical laws immediately involved, 'other things being equal.' But the removal of the latter condition means that to be sure of the correctness of the prediction, all other things that might be influential have to be taken into account. The conditions on the final state of the trajected penny range out in space and time to events, so to speak, within striking distance of the trajectory. There is evidently more than a single condition on the occurrence of the final state of the penny, but each such condition is itself multiply-conditioned. Thus, as we move back in a concrete retroduction of the event, we move back into a divergent series of conditioned events and processes. In the general case, the only condition one can place on this aggregate of conditions is that they be fulfilled. To demand further that they be not a coincidental aggregate but systematic would be to require that they be a very special case, that they share an orderliness such as that of the sequence of prior velocities and positions conditioning the earth's present velocity and position. Still, that demand is not a demand for a logical impossibility. One cannot exclude in some *a priori* fashion that the concrete pattern of diverging series of conditions be ultimately orderly. Our concern, however, is not with a logically-possible world, but with the world investigated by science, and within that context the hypothesis of ultimate systematizability is extremely doubtful. The understanding of the concrete pattern of diverging conditions is irremediably particular and concrete, whereas systematization is achieved by moving to the level of laws and primary relations and by abstracting from the empirical manifold.

The enormous aggregate of concrete divergent processes does not, however, fall into the category of the empirical residue. Coincidental aggregates of processes do not, indeed, come within the scope of statistical investigations, but coincidental aggregates of particular values do. How precisely statistical science yields general knowledge of aggregates of numbers has already been made plain in chapter two, and the value of such general knowledge will become clear in chapter five. It is worth noting here that the general and valuable knowledge given by statistics is in no way incompatible with full causal knowledge of the situation concerned. Thus one might know all the causes of growth and have all the details of height listed for a given population and still the validity and value of statistical understanding would remain. Again, the claim to the validity of statistical knowledge, as we have already pointed out, is not a claim to allegiance with the indeterminists. There need be nothing indeterminate about physical processes. It is classical science which seeks to explain such processes. The concern of statistical science is with numbers and distributions, and its unconcern for processes can hardly be taken as a criterion of the non-objectivity of those processes. The basic question of statistics is of the type 'how often does such-and-such occur?' or 'how many of such-and-such are there?,' where the such-and-such, the states, depend for their specification both on classical laws and definitions and on particular secondary determinations: e.g., 'height 6 feet,' 'mass 4,' 'energy $h\omega$.' The dependence of statistics on classical definitions and the distinction between the primary relativity of classical laws and the secondary determinations which are the object either of particular observations or of statistics are indeed but two of the factors relevant to current disputes about quantum theory. They are however factors which belong to statistical science in any form.

But one should note too how statistical science in general parallels classical science in its procedures. Thus, Reichenbach's third method of determining probabilities *a posteriori*, that in which a probability metric is inferred by means of general induction methods from known observational data,

> is carried through as follows: a probability metric is introduced in the sense of a hypothesis; then the observational consequences of the assumption are computed and tested, and thus the truth of the assumption is judged in terms of its consequences. The inductive inference is of the same type as any other inference from observational data to a hypothesis. The fact that in this case the hypothesis concerns a probability makes no difference.[45]

Just as a gas equation like $PV = C$ is an idealization which goes beyond experimentally correlated values for P and V, so a probability hypothesis is ideal both in its abstraction from randomness and in the classical definitions used.[46] The probability hypothesis is, however, more closely related to concrete results

---

[45] H. Reichenbach, *The Theory of Probability*, 360.

[46] The fact that science at any stage involves idealizations or abstract systems or conceptual schemes of this type is very important for discussions of scientific progress: cf. R. Harré, *Matter and Method*, on this point. The Von Neumann discussion of the irreplaceability of quantum theory is locked within the conceptual schemes of that theory and is thus questionable in its conclusions. Again, the ideal experiments related to the indeterminacy principle depend on the same scheme and so are conclusive only in relation to the scheme. This would seem to be the basis of M. Hesse's criticisms in chapter 10 of *Forces and Fields*, London, 1961. Cf. also B. Lonergan, *Insight*, 334–6; CWL 3, 358–360.

than its classical counterpart. This closeness may be gathered from the way in which the two types of abstraction occur in statistical science. The abstraction centres on numerical values and fractions, measurements subject to the influence of a divergent multitude of environmental factors; the positive aspect yields probabilities as cluster points of relative actual frequencies; the negative aspect consists in regarding random differences of relative actual frequencies from probabilities as nonsignificant. We will return to this question both in chapter six and in chapter eight.

Again, just as classical inquiry has to begin from data and search for the correct equations and functions so statistical inquiry, except in particular cases of antecedent symmetry,[47] begins from samples considered to be normal and searches for functions which will yield distributions of states. In both types of inquiry the process of curve fitting is relevant to the discovery of such suitable functions. To obtain these functions classical science normally draws on the store of differential equations and their general solutions, while statistical science, at least in the case of the quantum theory, uses operator equations which yield eigen-functions and eigenvalues, serving therefore both to select classes of events and to determine the respective probabilities of the selected classes. In both types of inquiry the functions of importance are normally continuous; the continuity, however, has a different significance in the two cases. Thus the continuous functions of classical theory refer to a continuum of events, the theoretical continuity of processes such as a planetary orbit. But in statistical theory the continuity of functions refers, not to a continuity of process but to the continuity of the ideal norm from which the observed values diverge at random.

This brief comparison of classical and statistical methods raises a variety of problems which will be dealt with in the following chapters. Thus, the aspect of negative abstraction in statistical science will be a focus of attention in chapter six; the process of verification and the problem of empirical reference will occupy us in chapters seven and eight; the manner in which classical and statistical methods complement one another will be treated in chapter five. Our immediate concern, however, will be the negative task of showing, as against Hempel, that statistical science is not a science of the singular: to this task we turn in the next chapter.

---

[47] H. Reichenbach, *The Theory of Probability*, 355–9.

Randomness, Statistics, and Emergence

In his *Probability, Statistics and Truth*, von Mises remarks: 'We have nothing to say about the chances of life and death of the individual, even if we know his condition of life and health in detail. The phrase "probability of death," when it refers to a single person, has no meaning at all for us.'[1] Why he adopts this position is patent from his treatment of statistical probability as being essentially concerned with collectives. That view has already been discussed to some extent in chapter two and some of its obscurities, which are considered objectionable, partly justified by an exposition of the peculiar type of insight involved in the foundations of a frequency theory such as that crystallized in von Mises' two basic axioms of randomness and convergence. Undoubtedly there is room for improvement and further clarification in von Mises' treatment:[2] but it is certainly true that any view of probability which does not include some form of von Mises' two basic axioms—which involve the notion of collective—will have excluded an element of meaning essential to probability theory as it is currently used in scientific practice.[3]

In contrast with von Mises' view of the inapplicability of statistics to the singular case there is Hempel's treatment of statistics which focuses attention precisely on the application of probability theory in the particular case.[4] That application, as Hempel admits, leads to certain problems and ambiguities. In order to be able to deal adequately with these difficulties and to expose precisely the flaws in Hempel's account of statistics it is necessary to consider at some length the more generic procedure of reasonable betting.

Hempel is by no means alone in his view of the applicability of probability theory to the particular case. Indeed, one of Kneale's main criticisms of von Mises' position is that it makes no allowance for application to the singular occurrence.[5] According to Kneale, use of probability rules in such cases is essential and obviously rational:

> Let us suppose that a man is considering whether to risk a stake *s* in the hope of a gain *g*. If he knows that the probability of his getting *g*

---

[1] Op. cit., 15.

[2] Cf. chapter 8.

[3] A large section of von Mises' book illustrates this, especially 174–220; cf. also Lindsay and Margenau, *Foundations of Physics*, chapters 5 and 6.

[4] 'Deductive-Nomological vs. Statistical Explanation,' *Minnesota Studies in the Philosophy of Science* III, Minneapolis, 1962, edited by H. Feigl and G. Maxwell, 98–169. Cited below as 'Hempel.'

[5] *Probability and Induction*, Oxford, Clarendon Press, 1949, 164–5.

by staking j is he can multiply *p* into *g* in order to get what is called the *mathematical expectation*. Whether or not it is rational to take the risk depends at least in part on a comparison of this with the stake. If the mathematical expectation *pg* is less than *s* which he risks, we say it is obviously irrational for him to undertake the venture.[6]

Again, the entire twelfth chapter of A. Pap's book, *An Introduction to the Philosophy of Science,* devoted to the range theory of probability, shows how closely this theory of probability is linked with the problem of singular prediction, and Pap himself would hold that

> if I were confronted with a box which I knew to contain coloured marbles, and I knew that each marble was either red or blue, then the question could be put in the form of a dichotomy: Is the marble that I am going to pick at random blue or not blue? Consequently, if I were forced to bet but given a choice of odds, it would certainly be rational for me to choose even odds.[7]

But why is it 'certainly rational,' 'obviously rational,' to bet in such a fashion? Braithwaite, in his 'Why is it reasonable to base a betting rate upon an estimate of chance?,'[8] in fact passes over this essential question and concentrates on mathematizing the process of lottery selection and on justifying the assumption of combinatorial invariance. Thus his final conclusion involves an assumption of rationality which parallels that of Pap and Kneale: 'An estimate of chance presupposes a belief in a permutational invariance. Once having accepted this powerful tool it is irrational not to make use of it again when it is needed for valuing a lottery and thus choosing a betting rate.'[9] The rationality which we wish to discuss is taken for granted.

Both Kneale and Reichenbach refer to the fact that 'most of our ordinary statements about chance are elliptical'[10] when referring to singular cases. Reichenbach goes on to expose his view of the meaning of such statements, which meaning is obviously related to the question of rationality:

> I regard the statement about the probability of the single case, not as having meaning of its own, but as representing an elliptic mode of speech. In order to acquire meaning, the statement must be translated

---

[6] Ibid., 120.

[7] *Introduction to the Philosophy of Science*, London, Eyre & Spottiswoode, 1963, 203; cf. also R. Carnap, *Logical Foundations of Probability*, University of Chicago Press, 1963, 226–8; 576–9.

[8] In *Logic, Methodology and Philosophy of Science*, edited by Y. Bar-Hillel, Amsterdam, North-Holland, 1965, 263–73.

[9] Ibid., 273.

[10] Kneale, *Probability and Induction*, 115; Reichenbach, *The Theory of Probability*, 376.

into a statement about a frequency in a sequence of repeated occurrences. The statement concerning the probability of the single case thus is given a *fictitious meaning*, constructed by a *transfer of meaning from the general to the particular case*. The adoption of the fictitious meaning is justifiable, not for cognitive reasons, but because it serves the purpose of action to deal with such statements as meaningful.[11]

Reichenbach's attempt to make precise the meaning of the 'singular' probability statement is not altogether successful. Firstly, he keeps reasonable betting too intimately connected with the frequency theory of probability. Secondly, even if it were true,[12] as von Neumann and Morgenstern would have it, that 'it may safely be stated that there exists, at present, no satisfactory treatment of the question of rational behaviour'[13] still there is the commonly accepted view that reasonable behaviour presupposes sufficiently sound reasons for behaving that way, and a fictitious transferred meaning which is unacceptable cognitionally will hardly 'serve the purposes of action.' The problem indeed is primarily cognitional: it is the problem of the reasonableness of expectation, whether a bet is involved or not. We will see that the need to act does in fact have a part to play and that Reichenbach's argument gives valuable indications of the direction of solution of the problem.

Let us pinpoint that problem. We wish to specify as fully as possible the meaning of 'because' or 'therefore' in the two following statements about a single draw from an urn containing three black balls to each white ball: 'It is reasonable to expect a black ball *because* there are three times as many black balls as white in the urn,' or, 'There are three times as many black balls, *therefore* it is reasonable to expect black.'

Obviously the ratio 3:1 is central to the meaning of 'because' here. Later we must make precise what is meant by 'expectation' but at all events it is clear that the expectation fluctuates with the ratio. We are not concerned immediately with the problem of measuring the expectation by the ratio: we merely note that the expectation would in some sense be greater if the ratio were, say, 100:1.

Now while one may say that there is some type of connection between the ratio and the expectation, there is a very definite lack of connection between the ratio or the expectation and the result of a single draw. The ratio is indeed in some way a reason for expecting a particular result. Central to this reason is an understanding of the structure of the draw-situation, and in the present case that understanding need make no appeal to statistical theory in von Mises' sense: it can be an understanding of an application of elementary combinatorial analysis.

---

[11] Reichenbach, op. cit., 376.

[12] Which is questionable: cf., e.g., B. Lonergan, *Insight*, 607–19; CWL 3, 631–642; F. E. Crowe, 'St Thomas and the Concrete Operabilia,' *Sciences Ecclesiastiques*, 1955.

[13] *Theory of Games and Economic Behaviour*, Princeton University Press, 1947, 9.

Similarly with an example involving the expectation of the next dice toss turning up a six: yet here we may note an important point which is not as obvious in the first example: the dice may be loaded; the white balls may be lighter, stickier, bigger, etc. In other cases one may, as Reichenbach would have it, avail oneself of a known statistical hypothesis: if, for example, we began to consider the odds of 20:1 on the next man we meet being 6′ 6″ tall—and even in this case, contrary to what Reichenbach says, our reasoning could in fact prescind from long-run considerations. After all, there may be only one bet, and the focus of our attention may not be on statistics.

But now let us pinpoint the problem further. Will the bet be reasonable? Or, to keep it more in cognitional terms, is the expectation e.g., of a black ball, reasonable? Certainly, if it is reasonable, it is reasonable in a peculiar sense, and that peculiarity is associated with the lack of connection already noted between the ratio or expectation and the result of a draw. Indeed in some sense the expectation of the drawing of a black ball is in no way a reasonable expectation: an understanding of possible outcomes or the ratio of types of possible outcomes, or the statistics of outcomes, cannot strictly be a reason for expecting in this case one outcome less than another, or even than many others. To be reasonable in the ordinary sense the understanding would have to include some understanding of the sequence of events leading to the outcome.

Can we possibly explain the rationality along the lines of von Wright?

> Let us assume, in predicting the event E n times under condition C, the prediction turned out to be true m times and false m' times, and that $m/n > m'/n$ ... We call it, for determinate reasons, *rational* to prefer the prediction of the event E under the circumstances C to the prediction of not-E ... We cannot with this assertion of "rationality" wish to exclude the possibility that, those circumstances being realized, the more probable prediction will after all turn out to be false ... We only demand that, if it is rational to prefer the prediction of E to the prediction of not-E, and it nevertheless happens that only a minority of actual predictions of E are true, then the distribution of truth and falsehood on the predictions must be regarded as representing a "chance-event" which cannot be excluded, but which we think, *in the long run,* will give place to another distribution where the true predictions are in the majority.[14]

von Wright's remarks help to expose further the 'twistedness' of the rationality of the betting process especially in noting that part of the rationality of the expectation includes the admission of the possibility of its falsification. He goes too far, however, in his attempt to 'rationalize' further the procedure. First of all,

---

[14] *The Logical Problem of Induction*, Oxford, Basil Blackwell, 1957, 142–3.

as we saw above, even if in some instances a frequency notion creeps in, it is not made use of in many cases. Again, the bettor is not always or even normally interested in holding his own in the long run and so it seems somewhat farfetched to build a long-run rationale into the procedure: the essential rationality of the betting process cannot be connected to some distribution of results, whether future or past, for its justification.

As the various authors point out[15] there is on occasion need for 'betting-action' not only among bettors but, for example, among medical men. The desire in either case is to come, in some sense, as near a prediction as possible. Moreover, except in the case of emotive illusion,[16] this falling short of prediction is appreciated as such: expectation is known not to be prediction, and the bettor may be disappointed but not startled if his expectation does not correspond to the outcome; if, for example, the white ball is drawn. The expectation differs from prediction in groundedness. This point is neatly put by G. B. Keene:

> If I am asked (or, for one reason or another want) to predict whether a particular event will occur, I normally look to see if there is any evidence pointing one way or the other, before making my prediction. If I can find none, my reaction would be "I just don't know" or "I wouldn't like to say." Sometimes, however, I *have* to say (as in calling while the penny is in the air), in which case I make a chance prediction ... what makes a prediction a chance prediction is its ungroundedness.[17]

But in the case of e.g., the urn draw, surely the expectation is grounded: is not this the nub of the argument of those who insist on the rationality of the expectation in the single case?

Consider the urn draw. The bettor knows that any one of four things may happen—we assume just four balls—but he lacks any further information about the events leading up to the drawing of the particular ball. If questioned about the outcome he would be reasonable in answering, as G. B. Keene remarked, 'I just don't know,' the reason being, to use Bernoulli's famous statement 'quia nulla perspicitur ratio cur haec vel ilia potius exire debeat quam quaelibet alia.'[18] We use that statement deliberately there as the ground for the claim 'I just don't know' in order to bring out a contrast with a more usual usage. For, Bernoulli's statement is the origin of the principle of indifference, or the principle of insufficient reason, where it now plays the role of ground for the determination of *a priori* probabilities. Now if the hypothetical bettor is interested in betting he

---

[15] Cf. Reichenbach, op. cit., 373; Kneale, op. cit., 117.

[16] Cf. Reichenbach, op. cit., 368.

[17] 'Randomness,' *Aristotelian Society, Supplementary Volume* (XXXI), 1957, 154.

[18] From *Ars Conjecturandi*, as quoted by Kneale, op. cit., 147. ['Because no reason is to be seen why this or that should be the outcome rather than any other one you care to mention.']

will in fact also appeal to this principle, implicitly yet genuinely. To use Kneale's phrase,[19] the principle is one of *absence of knowledge,* but the bettor's appeal to it in his expectation is grounded on a *knowledge of absence* of information. In ignorance of the outcome, and of the process leading to the outcome, the bettor expects with *equal* expectation each of the four results. The equality of the expectations thus in fact rests on an appreciation of a lack of knowledge coupled with a utilization of available knowledge. Accordingly, the procedure is reasonable in so far as the bettor has this appreciation of his weak position with regard to prediction and of the manner in which present information can be employed within that context when necessity demands.

This brings us an answer to the basic question, Is the expectation reasonable, and if so how is it reasonable? Firstly, in the absence of any need to bet or act, the reasonable attitude is expressed, not by an expression of expectation, but by an 'I just don't know.' But if a bet is required or desired, then the most reasonable, the best, bet is, in our example, on black—provided one is not giving odds of more than 3:1, and this proviso reveals something of the variability of reasonableness with changing context. It is the most reasonable bet because, while not grounded in G. B. Keene's sense, it has what might be called maximum available groundedness. And it is all the more subjectively reasonable in so far as the bettor appreciates intelligently the nature of the expectation and its strict lack of connection with the result: an appreciation normally revealed in his reaction to an unfavourable result.

In Reichenbach's terminology, the bettor should appreciate that 'the statement (e.g., "black will be drawn") is not an assertion but a posit: a posit is a statement with which we deal as true although the truth value is unknown.'[20] Moreover, the intelligent bettor

> endeavours to improve his posit by increasing the probability through a more precise analysis of the actual conditions, that is, by making a selection such that a greater probability will hold for the subsequence determined by s ... e.g. the physician may try to analyse the condition of his patient more exactly by taking X-rays.[21]

I.e. he aims at maximum possible, yet always essentially insufficient, groundedness, and he does this 'by considering the narrowest class for which reliable statistics can be compiled.'[22] There is, in fact, a stage in this narrowing when there may be a discontinuous transition from 'chance prediction' to 'grounded prediction,' when ignorance and *a priori* equiprobabilism etc. are

---

[19] *Probability and Induction,* 173.
[20] *The Theory of Probability,* 373.
[21] Ibid., 374.
[22] Ibid., 374.

replaced to some extent by causal knowledge of the particular involved, for example, in the case of the doctor treating a very familiar patient.

We began this discussion by noting the contrast between the views of von Mises and Hempel on the applicability of probability theory in the particular case. In the light of our considerations the weakness of Hempel's position and the source of what he calls the ambiguity of statistical explanation should become manifest. Put briefly and bluntly one might say that Hempel's analysis of statistical explanation centres on a process of ill-grounded or partially-grounded betting which is remote from strict scientific practice. Consider Hempel's general statistical syllogism:

> 'Almost all F are G,
> x is F
> ———————
> x is almost certain to be G.'[23]

Since it is general it should cover the example we have constantly been using; thus:

> Almost all balls are black (75 per cent in fact),
> x is a ball
> ———————
> x is almost certain to be black.

There is an obvious question here, whether Hempel would consider the conclusion to mean 75 per cent certain, but let us be satisfied for the moment with a more radical criticism. Following G. B. Keene's lead, we may say that the genuine logic here is:

> Almost all balls are black
> x is a ball
> ———————
> I just don't know if x is black.

Thus Hempel's conclusion simply does not follow scientifically. To make it follow at all one has to move away from statistics as it is used in science, and in the general case—when the major premise is in fact an *a posteriori* frequency-statistical hypothesis—one may be using such statistics to come to an *expectation* in the manner already described. The conclusion then should be: x may be expected to be black. At all events the 'almost all' of the major premise cannot reasonably be slipped into the conclusion as a qualification of certainty.[24] Indeed one might say that expectation has nothing to do with certainty, with prediction,

---

[23] Hempel, 125.

[24] Here one may notice that the probability notion in the conclusion is the 'degree of confirmation' one: implicitly too we are criticising various 'induction by enumeration' views, Laplace's $(n + 1)/(n + 2)$ rule, and Carnap's more elaborate version of it (*Logical Foundations of Probability*, 226–228, 567–9). Cf. the conclusion to the present chapter.

with prior knowledge: rather does expectation of this type imply grounds for uncertainty and for judgment of uncertainty, 'I just don't know nor can I know.'

In his conclusion,[25] Hempel remarks on the difficult variations of the meaning of 'because' in statistical as opposed to deductive-nomological explanation, and it seems that the entire structure of his investigation might have been modified had he analysed initially the meaning of 'because' in such statements as '*because* three out of four balls are black, the chance of the next ball picked being black is 3/4.' With the failure to reach the meaning of expressions such as this is associated that of not appreciating the essential role of coincidental aggregates in statistics: 'there is only a difference in degree between a sample consisting of just one case and a sample consisting of many cases.'[26]

The example we took was a simple one of expectation based on combinatorial analysis and it illustrated clearly the impossibility of grounded prediction in cases of this type. More complex cases, however, are liable to involve not only statistical hypotheses or the like, but information and science which could in fact contribute to a grounded prediction. For that reason one must always distinguish the elements within the grounds of an expectation or a prediction. At one extreme there is the ordinary grounded prediction, at the other there is the sufficient groundedness, based only on statistics or such like, for a reasonable bet. But between these two there can be a range involving some mixture of pure betting expectation and grounded prediction, and the transition from one extreme to the other can sometimes be accomplished by what Reichenbach calls a narrowing of classes.

But lest we seem to be unfair to Hempel in using our example of his syllogism, let us consider this own illustration and go on to resolve his 'ambiguity of statistical explanation.'

> Consider the following argument which represents, in a nutshell, an attempt at a statistical explanation of a particular event: John Jones was almost certain to recover quickly from his streptococcus infection, for he was given penicillin, and almost all cases of streptococcus infection clear up quickly upon administration of penicillin. The second statement in the explanans is evidently a statistical generalization and while the probability value is not specified numerically, the words "almost all cases" indicate this is very high.[27]

Hempel's example fares no better than our own, when considered in the light of our earlier discussion. The doctor reasonably expects John Jones to recover, and if he does not recover he suspects that there were added compli-

---

[25] Hempel, 166–7.

[26] Ibid., 132.

[27] Ibid., 125.

cations, but he is certainly not amazed in this latter case, since not everyone recovers, and *as far as he knew* there was nothing about John Jones to put him either in the majority group or in the unfortunate minority. The significance of all this is clear from our previous consideration of the reasonableness of expectation in the general case.

But now—and this brings us to the ambiguity problem—let us suppose that John Jones had peculiarities of, for example, blood circulation, and that people with such peculiarities rarely recovered from streptococcus infection, with or without penicillin. Then the doctor, *had he known this,* might have argued thus: 'Almost all people with streptococcus infection and blood peculiarities of this type die, with or without penicillin; John Jones is one of these; therefore I would not expect him to recover.' But of course the doctor might add, to Mrs. Jones, 'he may indeed recover.' Hempel, with a similar example, comes up with a conclusion like ours which contradicts the first conclusion. In the first case, 'John Jones is almost certain to recover'; in the second case, 'John Jones is almost certain not to recover.' Thus Hempel comes to speak of 'the ambiguity of statistical prediction and, more exclusively, of the ambiguity of statistical systematization.'[28] On our view, there is no question here of prediction, nor is Hempel's syllogism characteristic of statistical systematization or its use.[29] It is to be admitted, however, that there is indeed room for ambiguity within the process he describes, the process of reasonable betting. This latter point fits easily into the context of our justification of the 'reasonableness' of expectation. Expectation is based on knowledge of absence of information and use of available information. The maximum available groundedness is had for an expectation when all available information is correctly used. But there can be a change in the available information which could reverse the expectation, or there may even be alternate ways of narrowing the classes involved[30] which lead to ambiguities. All these can, however, be sorted out and appreciated by the person interested in *expecting* or betting in so far as he knows what precisely is involved in the process, so that the ambiguity is reduced to an appearance of ambiguity or at least clearly exposed. So, while we may modestly agree with Reichenbach in stating that 'we do not affirm that this method is perfectly unambiguous,'[31] we see no reason why the ambiguity should not be rendered harmless.

From the point of view of strict scientific practice the ambiguity is rendered harmless by the simple exclusion of the application of probability to the particular case. The following chapter illustrates the procedure and the consistent implicit exclusion of such application. But to lay the axe to the root of the problems and

---

[28] Ibid., 127.

[29] Cf. the following chapter, towards the end.

[30] H. Reichenbach, op. cit., 374.

[31] Ibid., 375.

ambiguities underlying application of statistics to the singular instance requires a more refined analysis which will be undertaken only in a later context. For, underlying the difficulties of Hempel's analysis there is the problem of two distinct notions of probability, one of which is associated with the statistical hypothesis, the other with confirmation or verification or prediction. Most of Hempel's illustrative syllogisms indeed involve an illegitimate switch from hypothesis-probability in the major premise to the confirmation notion of probability in the conclusion. Furthermore, there is the related problem raised by the question, of what is probability a property? A large group of authors would answer that, for example, the ½ is a property or disposition of a penny, or at least of a penny in a given environment: 'a probability₂ value is a physical property like a temperature.'[32] 'To assert that the probability of a normal coin presenting heads after being tossed is ½, is to ascribe a physical property to a coin which is manifested under determinate conditions.'[33] 'Since the probabilities turn out to depend upon the experimental arrangement, they may be looked upon as *properties of the arrangement.* They characterize *the disposition or the propensity* of the experimental arrangement to give rise to certain characteristic frequencies *when the experiment is often repeated.*'[34] These views on probability as a property, with which we do not agree, are related to basic assumptions about meaning and verification. Hence both this problem and that of the two notions of probability point to the need for a discussion of verification for their adequate solution. That discussion and adequate solution will concern us in chapter seven.

---

[32] R. Carnap, 'Two Concepts of Probability,' *Readings in Philosophical Analysis*, edited by Feigl and Sellars, New York, Appleton-Century-Crofts, 1949, 346. Probability₂ refers to the frequency notion of probability.

[33] E. Nagel, 'Principles of the Theory of Probability,' *International Encyclopedia of Unified Science*, Vol. I, pt. 2, 365. Cf. also Hempel, 123.

[34] K. Popper, in *Observation and Interpretation in the Philosophy of Physics* edited by S. Körner, New York, Dover, 1958, 67.

The Method of Residues is one of the most important among our instruments of discovery. Of all the methods of investigating laws of nature, this is the most fertile in unexpected results: often informing us of sequences in which neither the cause nor the effect were sufficiently conspicuous to attract of themselves attention of observers.[1]

Such is Mill's comment on his fourth method of experimental inquiry. Our aim in the present section is to justify Mill's claim by specifying the types of residue involved in empirical investigations and by showing their significance for the progress of inquiry. In particular our interest centres on statistical residues, the existence of which, indeed, Mill would hardly grant: further on he excludes chance, discussing there the discovery of residual phenomena by the elimination of the effects of chance. Such a view is to be expected, given his position on abstraction,[2] one which we have criticized generically already.[3] For this reason, and also because statistical methods were not highly developed in his day, it is no cause for wonder that none of Mill's Canons of Inquiry touched on the techniques of statistical science. His method of residue is, then, in need of essential amplification. He formulated this method in his fourth canon: 'Subduct from any phenomenon such part as is known by previous induction to be the effect of certain antecedents, and the residue of the phenomenon is the effect of the remaining antecedents.'[4] Even taken in the limited sense that Mill intended the Canon has all the power that Mill claimed for it. He illustrates that power with examples from Herschel's 'Outlines of Astronomy,' and it is worth noting Herschel's prior appreciation of the value of the method in astronomy:

Almost all the greatest discoveries in Astronomy have resulted from the consideration of residual phenomena of a quantitative or numerical kind … It was thus that the grand discovery of the precession of the equinoxes resulted as a residual phenomenon from the imperfect explanation of the return of the seasons by the return of the sun to the same apparent place among the fixed stars.[5]

---

[1] J. S. Mill, *A System of Logic*, 260.

[2] Ibid., 424–433; cf. also J. Mill, *Analysis of the Phenomena of the Human Mind*, London, Longmans, Green, Reader & Dyer, 1869, edited by J. S. Mill, Vol. I, 233–7; 247–317.

[3] Above, 35 ff.

[4] *A System of Logic*, 260.

[5] *Outlines of Astronomy*, 1856, quoted by Mill, *A System of Logic*, 281.

The residual phenomena which Herschel and Mill mention are phenomena which are not accounted for by a particular explanation but whose existence becomes manifest through the use of that explanation. In contrast with this type of residue there is the empirical residue which we introduced earlier.[6] That residue is, one might say, a residual phenomenon which is part of the given yet which calls for no further explanation. Our interest in the present section centres on another type of residue, the statistical residue. Most of our efforts so far have in fact been directed to showing that there are statistical residues as well as the merely empirical residue: to show, in other words, that when all classical laws are understood there is left over not only the residue which no one bothers to explain, the empirical residue, but a residue unsystematizable by classical laws which consists in, for example, the coincidental aggregate of velocities of molecules in a gas at a given instant, or the aggregate of heights in a population. This residue, unlike the empirical residue, is not entirely opaque: with it is associated the understanding gained through statistical methods—whence its name.

To appreciate the role and significance of such residues in scientific investigations we had best turn to the examination of particular instances of inquiry. We are interested especially in exposing the types of anticipation which correspond to statistical residues, where by 'anticipation' we mean something closely resembling Churchman's 'presuppositions' where 'the term "presuppositions" is used here, not in the vague sense of the *a priori* in metaphysical literature but in the technical sense used above: a proposition which (implicitly or explicitly) is asserted in all the acceptable alternative responses to a question.'[7] Our investigations so far have enabled us to appreciate as the central anticipation of statistical method the anticipation of randomness in the object of investigation: '. . . randomness really dictates the method we are to use in making predictions, and if randomness does not exist, then statistical techniques of handling data are not in general applicable.'[8] Such an anticipation need not be an explicit proposition in the investigator's mind: it is the methodologist's task to formulate and criticize it. In the mind of the investigator it is normally a nonpropositional anticipation, a habit of inquiry, a habit which is *a posteriori*, which grows and becomes more refined through scientific practice and scientific advance. Aristotle, for example, had no anticipation of statistical explanation, although he acknowledged the existence of residues. The only laws known to him were primitive versions of classical laws with their associated models: besides these there was the contingent, the coincidental—probability theory had to await the gambling problems of the seventeenth century. Nowadays the anticipation of

---

[6] Above, 22, 37.

[7] C. West Churchman, *Theory of Experimental Inferences*, New York, Macmillan, 1948, 42.

[8] Ibid., 6.

statistical explanation is part of the working habits of any practicing scientist. Such an anticipation is in no sense an *a priori* bias in investigation, for the scientist has other anticipations, such as the basic anticipation of classical explanation, an anticipation of system. Faced with a particular problem he may formulate a hypothesis on either anticipation and try it out. It is worth noting that here an aspect of the complementary of classical and statistical investigation is evident: for one may anticipate either that the data is systematic or that it is non-systematic. The point is neatly illustrated by Keene's example of a situation in which there is a possibility of a systematic factor entering in, in a penny tossing experiment.

> The first thing we should do, in this case, would be to examine the penny very carefully to see if there was any reason why it should fall one way rather than the other. Suppose we could find no reason. Then there is a choice open to us—namely between accepting the evidence for its physical symmetry of the penny and accepting the evidence for its physical asymmetry. If we accept the evidence for asymmetry we reject the claim that the penny is a randomizer. That is, we regard ourselves as being able to make grounded predictions. If we accept the evidence for symmetry, on the other hand, we, *ipso facto,* rule out the possibility of any claim to be able to make grounded predictions.[9]

The example illustrates in the simplest way the oscillations of attitude within the process of inquiry. But let us appreciate these oscillations, and much else besides, in significant scientific work.

The experiments, which we will discuss in some detail, have to do with the distribution, survival and more precise definition of buttercups.[10] There are three closely related species of the genus *Ranunculus*, the buttercups *Ranunculus bulbosus*, *Ranunculus acris* and *Ranunculus repens*. All three species are inhabitants of much the same type of community, unshaded herbaceous grassland. These experiments fall into four main categories, the first two of which will serve as our illustrations: (a) detailed investigation by statistical methods of the nature of buttercup distribution in permanent pastures and hay meadows, (b) transplant and sowing experiments with the three species in pots under controlled moisture conditions.

The first group of experiments was concentrated on two areas of permanent grassland, Port Meadow and Pixey Mead, near Oxford, which have been continuously grazed and mown respectively for at least eight hundred years. Both areas were sampled by throwing 10″ square quadrats at random in an area chosen for its uniformity of topography and vegetation. On Port Meadow *Ranunculus*

---

[9] *Aristotelian Society Supp. Vol.* XXXI, 1957, 'Randomness,' 158–9.

[10] J. L. Harper and G. R. Sager, 'Some aspects of the ecology of buttercups in permanent grassland' *Proc. British Weed Control Conf.* (I) 1953, 256–63.

bulbosus was the only abundant buttercup. The distribution of the species was tested for randomness by the percentage absence method and also by the ratio of variance to mean. The distribution of both mature plants and seedlings (it was spring) was markedly nonrandom. On Pixey Mead, where *Ranunculus acris* and *Ranunculus repens* were abundant the apparent uniformity of distribution of the plants was a misleading visual impression and there was a marked heterogeneity shown by a significant nonrandomness in the distribution. The nonrandomness of the distribution of all three species was due to clumping.

Let us pause to take stock of the assumptions and anticipations so far involved. The first assumption to be noted is that of the available definitions of the three types of buttercup: this assumption illustrates a point noted already— the dependence of statistical method on classical science for its definitions. Again, this initial stage of the investigation is governed by a statistical anticipation. One characteristic of this anticipation to be noted here is that it does not look to, anticipate, a redefinition of the classes or states involved, but as we shall see it can mediate such a change. There is, furthermore, an initial elementary statistical assumption, or hypothesis of randomness: we pass over the more fundamental assumption of empirical residue connected with the definitional nonsignificance of place in the particular fields. This hypothesis is in fact falsified, but in the process of falsification a clue to further hypothesis is made available. That clue might have been formulated more precisely as a new statistical hypothesis giving mathematical form to the generic description of 'clumping': but such an effort at specification is uncalled for. The generic conclusion from the numerical data to clumping is sufficient to point to some type of classical hypothesis, to something relating to a systematic differentiation of places.

Several such hypotheses are available. Firstly, clumping is commonly associated with species which multiply vegetatively: there is clearly a classical correlation involved here. But this hypothesis is quickly falsified in the present case, for *Ranunculus bulbosus* scarcely ever multiplies vegetatively in the field. A second hypothesis would put the clumping down to a strong tendency for seedlings to grow up in colonies around parent plants. This hypothesis was in fact supported by an analysis of the frequency of association of plants and seedlings in the random quadrats. Here we have a statistical correlation hypothesis, and it would seem that by using the previous data—somewhat along the lines of Mill's method of residues—one could have determined the presence of a residual phenomenon and the need for a further hypothesis. The paper under discussion, however, introduces this further, third, hypothesis more directly: it is the hypothesis that clumping might be due to differences in microhabitat within the communities sampled. Note here an anticipation of the classical type: one is proposing a systematic factor, some type of correlation of plants with specific microhabitats. To get clues to a specification of plant-habitat relations, was recourse had to statistical methods. An analysis was made of the distribution of buttercup species on ridge and furrow grassland: here a convenient pattern of variation in microenvironment is imposed on the area by the undulations of the land surface. The results are shown graphically in Figure 1 (see diagram p. 60). *Ranunculus bulbosus* was found to occupy bands of land running along the tops of ridges; *Ranunculus acris* occupied the sides of the ridges and *Ranunculus repens* lay in the furrows. Omitting details, it was clear that the three species could be arranged in an order correlated with field drainage, with water table. Note here how the statistical method leads gradually towards a redefinition, or at least a more ample definition, of the classes with which it deals. To make more precise the correlation of plants and water tables the second category of controlled experiments was undertaken, where growth, flowering, etc., of the plants were observed in pots with controlled water tables. Figure 2 (see diagram p. 61) gives some indication of the direction of such experiments. But we have perhaps already sufficient illustration of the way in which the two procedures, classical and statistical, complement one another.

Figure 1

THE DISTRIBUTION OF THREE SPECIES OF RANUNCULUS ON RIDGE AND
FURROW GRASSLAND

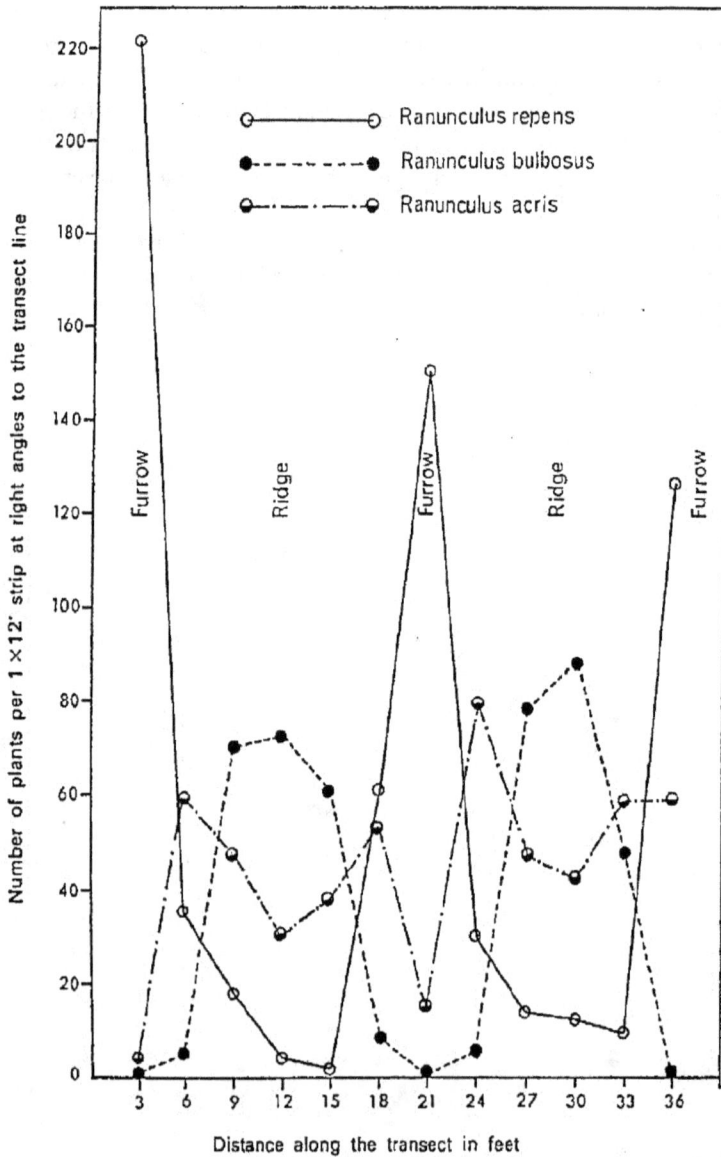

Figure 2

THE EARLY ESTABLISHMENT OF THREE SPECIES OF RANUNCULUS SOWN IN
POTS WITH THE WATER MAINTAINED AT DIFFERENT LEVELS

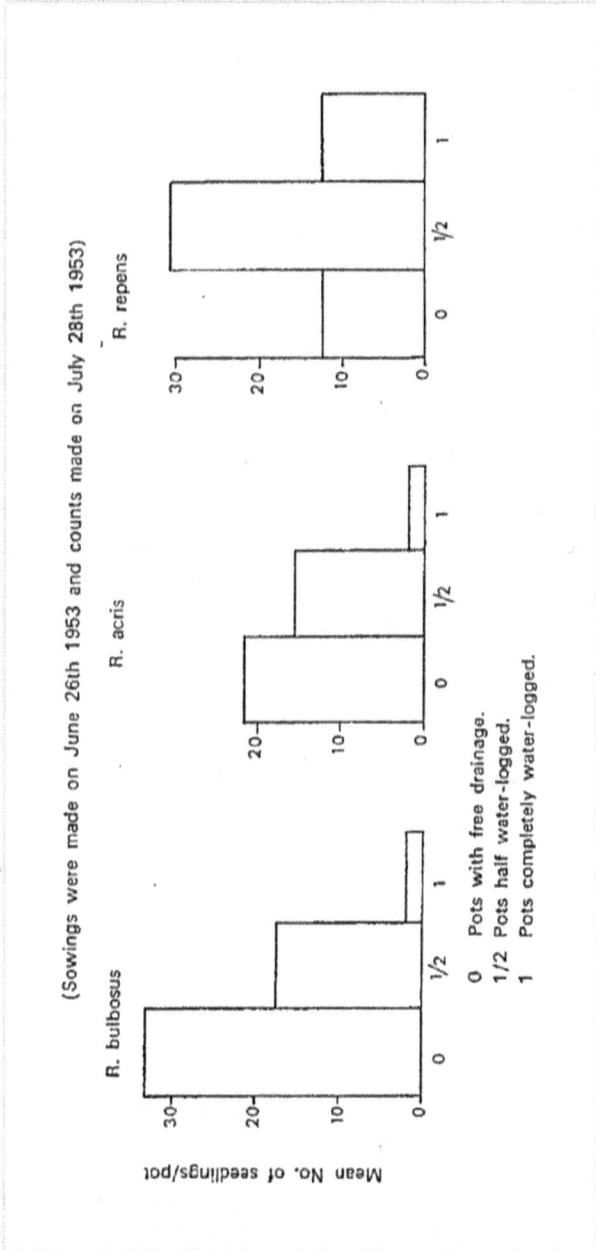

The second set of experiments illustrates clearly the principle of physical separation and controlled experimentation, but the first set illustrates better the less obvious principle of mental separation which is continually operative in science. Both principles are implicit in Mill's canons with reference to classical laws, and, with reference to statistical laws, in Yule and Kendall's remark:

> Experiment seeks to disentangle a complex of causes by removing all but one of them, or rather by concentrating on the study of one and reducing the others, as far as possible, to a comparatively small residuum. Statistics, denied this resource, must accept for analysis data subject to the influence of a host of causes, and must try to discover from the data themselves what causes are the important ones and how much of the observed effect is due to the operation of each.[11]

The mental separation is normally carried out by using one's previously acquired knowledge of laws that are, or may be, involved in order, so to speak, to close in on the unknown. Such a process was in operation in the consideration of the possible reasons for clumping: for example, knowledge of the non-vegetative multiplication of *Ranunculus bulbosus* excludes one hypothesis. Where physical separation is impossible, usually because it would disturb the data in essential features, this process becomes necessary. This is the case in much of meteorology, anthropology, sociology '. . . the observer of social facts cannot experiment, but must deal with circumstances as they occur, apart from his control.'[12] Again, it is true of experimental work on living organisms, where the mental separation is often complemented by the use of 'control' organisms, and where 'even the most skillful and critical workers are to some extent frustrated in performance of the ideal experiment by the very characteristics of the organism which they attempt to understand.'[13]

Most of Mill's illustrations are drawn from Astronomy[14] and they could be complemented by illustrations from modern perturbation theory, both classical and quantum.[15] A simple instance of this mental separation in chemistry is the manner in which Boyle's law is used in determining Charles' law: the condition that the pressure remain constant which occurs in Charles' law would otherwise lack significance. Again, both Boyle's and Charles' law are assumed in the determination of Gay-Lussac's law of volumes, which holds 'if all measurements

---

[11] Yule and Kendal, *Introduction to the Theory of Statistics*, xv.

[12] Ibid.

[13] L. J. Barth, *Development: Selected Topics*, Reading, MA, Addison-Wesley, 1964, 64.

[14] *A System of Logic*, 280–282.

[15] For example, Heitler, *The Quantum Theory of Radiation*, 3rd edit., Oxford University Press, 1954, section IV; some interesting doubts regarding the type of 'mental separation' involved in such quantum mechanical problems are raised by Schrödinger, *British Journal for the Philosophy of Science*, 1952–3, 'Are there quantum jumps?,' 121–2.

are made under the same conditions of temperature and pressure'[16] and which could be verified even in the absence of these conditions through the use of the other two laws. Nearer to our subject—for it is a case of mental separation through statistical laws—is the use of the law of probable errors in excluding non-systematic deviations from experimental results. [17] And perhaps our entire discussion is best illustrated by the famous experiments and conclusions of Mendel. So, for example, the problem of abstraction from non-systematic divergences to an ideal norm is neatly put in Woodger's comment on Mendel's conclusion from his count in the pea experiment:

> So long as we assert that the average ratio is 2.98 to 1 we are dealing with accessible sets and have no law or explanatory hypothesis. But what does Mendel's addition "or 3 to 1" mean? Presumably these words express the leap from an observed proportion in an accessible set to a hypothetical proportion in an inaccessible set.[18]

The appearance of different types of pea plants was non-systematic. Mendel's conclusion '3 to 1' brought sufficient statistical generalization to lead him to the classical-type postulate of dominant and recessive genes. The non-systematic phenomena were then explained by statistical combinations of classically-defined genes.

We have illustrated the way in which the two types of investigation, classical and statistical, complement each other by anticipating respectively system and absence of system in the object of inquiry, and by thus helping towards a separation, mental if not physical, of one from the other. But our illustrations also show that both types of inquiry anticipate system: for the result of each type of investigation is some type of systematic or general hypothesis. There is no contradiction in this: it merely brings out two aspects of the anticipation in question: anticipation is of a type of object to be investigated, yet is also of the result of investigating that object.[19] In the case of classical laws, one anticipates, in this latter sense, a systematic hypothesis—for example, Newton's inverse square law defining masses and the resulting orbits—on which actual results are expected to converge. In the statistical case one seeks a systematic hypothesis, like the normal distribution, from which particular results are expected to diverge,

---

[16] Cf. Any standard text book, for example, Durrant, *General and Inorganic Chemistry*, London, Longmans, 1964, 12.

[17] An interesting question in this connection is whether something parallel to this may not evolve in quantum theory, opening the way to further systematization at the subatomic level: cf. Lonergan, *Insight*, 107; CWL 3, 130.

[18] J. H. Woodger, 'Studies in the Foundations of Genetics' in *The Axiomatic Method*, edited by Henkin, Suppes and Tarski, Amsterdam, North-Holland, 1959, 415.

[19] The distinction is found in Aquinas: for references and discussion cf. Lonergan, *De Deo Trino*, Vol. II, Rome, 1964, 273–280; CWL 12, 569–575.

but not systematically. This last statement has a paradoxical ring about it, but it describes neatly the spontaneous anticipations of the statistical investigator. An experimenter would be surprised if he found that a sample fitted perfectly to a particular normal distribution curve: he expects non-systematic divergence. 'The theory of random sampling leads us to expect a certain measure of departure ... if this departure is not realized, something is wrong.'[20] But if the departure is not random but systematic the investigator is led to question not only his hypothesis but also his underlying assumptions. Such a revelation of systematic divergence would correspond to Mill's 'discovery of a residual phenomenon eliminating the effects of chance' which he illustrates with the example of the loaded dice:

> If the dice were not loaded, and the throw were left to depend entirely on the changeable causes, these in a sufficient number of instances would balance one another, and there would be no preponderant number of throws of any one kind. If, therefore, after such a number of trials that no further increase of their number has any material effect upon the average, we find a preponderance in favour of a particular throw, we may conclude with assurance that there is some constant cause acting in favour of the throw.[21]

This is as near as Mill comes to the principle of oscillation between statistical and classical explanations or anticipations.

The same point is illustrated by the first set of buttercup experiments. Here there was an initial hypothesis of a uniform distribution in the fields. 'Uniform distribution' is, of course, a systematic hypothesis concerning the secondary determinations of position of plants in the fields. Non-systematic divergence from uniform distribution is expected and allowed for. But the results showed on analysis a systematic divergence, through clumping, and the need for a further hypothesis. This further hypothesis could be an improved statistical hypothesis, or more directly a 'leap' to a classical correlation of plant and water table.

The statistical and classical hypotheses not only grow from, or build on, each other, but they depend on each other in formulation. The classical hypothesis and its predictions are always conditioned by the assumption 'other things being equal' and this, as we saw earlier, is a reference to the complex of conditions where we have found statistical residues. On the other hand, the statistical hypothesis is always dependent on the classical laws for its definitions. This is very evident in the examples so far given: the distributions, say, of buttercups or peas, are of sufficiently well-defined classes, and the definitions do not come from statistics itself. This is noted in a significant way by Yule and Kendal:

---

[20] K. Mather, *Statistical Analysis in Biology*, London, Methuen, 1965, 23.
[21] *A System of Logic*, 349–50.

> The methods of statistics ... deal with quantitative data alone. The quantitative character may, however, arise in two different ways. In the first place, the observer may note the *presence* or *absence* of some attribute in a series of objects or individuals and count how many do or do not possess it ... The quantitative character, in such cases, arises solely in the counting. In the second place, the observer may note or measure the actual magnitude of some variable character ... for instance, the ages of persons at death ... The observations in these cases are quantitative *ab initio*.[22]

The first way mentioned is very directly connected with the empirical residue discussed earlier: thus, the attributes may be verified at different places and different times, but these differences do not add to the definition of the attributes, nor, inversely, do the attributes involve any reference to particular locations and times. Here too we have another aspect of complementarity: the definition of types of events and things falls to classical science; their number and distribution can be known only by particular observations or in general by statistical laws. 'The law of the lever tells us nothing about the frequency of levers, about the places where they are to be found, about the times at which they function.'[23] The distribution of *Ranunculus bulbosus* is a clear example of this first way. It is also exemplified by the occupancy problem, say the distribution of r objects in n cells, which varies interestingly according to the conditions of indistinguishability or occupation limitation imposed.[24]

The second way mentioned has reference to the secondary determinations as opposed to the primary relativity that is given by the classical correlation. Thus, Newton's inverse square law and the general conic equation derivable by integration give the primary relativity between masses: but particular values are needed to have the secondary determinations, to know the precise ratio of masses, etc. Illustrative of this second way are the distributions of heights and of energies, or scattering distributions.

In the light of all this it should be clear that statistical laws are of no greater scientific significance than the definitions of the events whose frequencies they give. So one may conclude that the hope that statistical science will someday oust classical science is a hope for the triviality of a statistics of undefined events: trivial because the undefined event occurs all the time everywhere. Again one may conclude that statistical techniques should not be allowed exaggerated significance or complexity in relation to available classical definition: thus, for example,

---

[22] Yule and Kendal, *Introduction to the Theory of Statistics*, 1.

[23] Lonergan, *Insight*, 113; CWL 3, 136.

[24] Cf. W. Feller, *An Introduction to Probability Theory and Its Applications*, New York, Wiley, 1950, Vol. I, ch. 3.

the application of the powerful statistical theory of stochastic processes to historical development seems, to say the least, premature.[25]

One may well ask whether quantum mechanics is not an exception to this claim regarding the restriction of a statistical theory by a complementary classical theory through its dependence on it for definitions. We may fruitfully put the answer to the question in the context of Hanson's discussion of Correspondence and Uncertainty.[26] In that discussion Hanson points out that 'the correspondence principle and the uncertainty principle are inevitably in conceptual conflict';[27] 'first we are warned that the new physics is logically different from the old, and that we should not make old-fashioned demands on it. Then we are told that the two are conceptually quite harmonious. This needs sorting out.'[28] Our comments below are a contribution to that sorting out: but it must be emphasized that we are interested rather in showing the generality of our account of the interplay of classical and statistical science, and its relevance to problems of this type. To bring it to bear in detail upon this problem would be a much longer task.

First of all, then, we note that what underlies the correspondence principle is in fact the general thesis concerning the dependence of statistical science on classical science for its definitions. As can be seen in any textbook on the subject, quantum physics concerns itself with such things as energy, angular momentum, etc., and the relations between these theoretic elements are drawn from classical physics.[29] As in any statistical science, that concern is directed to particular values, to the secondary numerical determinations of the primary relations given in the classical definitions and to their distributions. We pointed out already, however, that quantum mechanics is peculiar in that its operator equations yield both particular values and distribution functions together. Briefly then, classical physics provides a correlational definition of energy, quantum physics may give a spectrum of energy values with associated probability distributions.

Next let us recall the correspondence rule as expressed simply and relatively by Heitler:

> Classical mechanics is contained in quantum mechanics as a *special case*. If we consider the behaviour of particles with heavier and heavier masses according to quantum mechanics, it turns out that, owing to the small value of Plank's constant, all probability distributions contract into almost certainty. It is then possible to assign to both position

---

[25] Cf. *Philosophy of Science*, 1960, 'On the Analysis of History and the Interdependence of the Social Sciences,' F. M. Fisher, 147–158.

[26] N. R. Hanson, *The Concept of the Positron*, Cambridge University Press, 1963, ch. 4.

[27] Ibid., 60.

[28] Ibid., 63.

[29] Cf. for example A. Messiah, *Quantum Mechanics*, Amsterdam, North-Holland, 1964, Vol. I, 29–34; 68–71; 216–218, etc.

and velocity almost sharp values and the behaviour of such bodies is one of near determinism. This amounts to practically complete determinism when the masses are as big as say those of a dust particle.[30]

We recall immediately, in regard to the assignment of sharp values to position and velocity, that classical laws are abstract, that, like point and line in geometry, position and velocity are defined implicitly within the abstract system. Furthermore, the definition of velocity is by the abstractive process of taking a limit, and it is interesting to note that this particular abstractive process is invariably involved in one of the correlated or conjugate variables. The understanding, then, which classical mechanics—Newton's or Einstein's—gives of the motion of particles is ideal, like the understanding of the ideal gas, or of the continuous fluids of hydrodynamics. Part of that ideal is the possession of an instantaneous determinate velocity at a determinate position. As with any classical law, particular experimental results converge on that ideal: whence Heitler's remark: '. . . the behaviour of such bodies is one of near determinism.' Now on the micro-level that experimental convergence is cut off because of the fact that observation entails a nonzero interference with the observed. Furthermore, we recall our specification above of the concern of quantum mechanics: as Messiah remarks, 'all the conclusions of the Quantum Theory can be put into the following form: "One obtains this or that result if one makes this or that observation."'[31] G. W. Mackey, too, associates observables with a particular type of question '. . . it corresponds to asking the (yes or no) question: Did the measurement of A lead to a value in E?'[32] In chapter seven we will deal further with the types of question, when we treat of Aristotle's divisions of questions, in the context of the problem of verification. In present terms, quantum mechanics is a theory concerned with the explanation and prediction of secondary determinations, the primary correlations being given by classical physics. This being the case, if the theory is not to overpredict, its axiomatic formulation must contain an axiom which suitably cuts off prediction: an axiom of indeterminacy, limiting the prediction of values of conjugate variables in a manner to be determined by investigations of the minimum of experimental interference. Note finally that such investigations are always carried out within, and thus limited by, the framework of present conceptual schemes.

Such a view of the present structure of quantum theory, in contrast with the dogma of Indeterminism, is a prosaic *a posteriori* conclusion based on a general analysis of the nature of statistical theories and their dependence on classical theories. Quantum theory is here taken to be a statistical theory limited

---

[30] W. Heitler, 'The Departure from Classical Thought in Modern Physics,' *Albert Einstein: Philosopher-Scientist*, ed. P. A. Schilpp, New York, Tudor, 1949, 195.

[31] Messiah, *Quantum Mechanics*, Vol. I, 151.

[32] *Mathematical Foundations of Quantum Mechanics*, New York, 1963, 64.

conceptually by the complementary classical theory, and if, as Einstein remarked,[33] there is no room in the conceptual world of statistical quantum theory for a complete description of the individual system, it is among other things, because there is no room for it within the conceptual world of present classical theory. One is reminded of Schrödinger's comment on quantum jumps:

> There have been ingenious constructs of the human mind that gave an exceedingly accurate description of observed facts and have yet lost all interest except to historians. I am thinking of the theory of epicycles. I confess to the heretical view that their modern counterpart in physical theory are the quantum jumps.[34]

But whatever about future physics, we have perhaps succeeded in showing the relevance to the problem of present physics, which Hanson points out, of our general discussion. The Correspondence Principle is to be associated with definitional continuity of classical and statistical physics, and so with the primary correlations given by definition; the principle of indeterminacy is to be associated with secondary particular determinations of these relations and the experimental limitation of their determination.

Finally we may add that the word 'complementarity' as used here has nothing to do with Bohr's principle of complementarity. Bohr's principle is linked with physics and has its origin in the fact that pairs of conjugate variables are normally associated, one with periodic, one with nonperiodic properties of microentities. There is room for complementarity in our sense in all the sciences: it is a complementarity of statistical and classical explanations which aids towards the goal of complete explanation. There are, however, other views of complementary classical models for the same phenomenon, views which are in fact a hindrance to scientific inquiry.[35]

Our discussion of the dependence of statistics on classical science throws a certain amount of light too on the problem of the relation of classical to statistical thermodynamics. First we may note that the customary opposition of

---

[33] *Albert Einstein: Philosopher-Scientist*, Einstein's 'Reply to Criticisms,' 668.

[34] E. Schrödinger, 'Are there quantum jumps?' *British Journal for the Philosophy of Science* (1952–3), 112.

[35] For a criticism of the view of complementarity mentioned, cf. R. Harré, *Theories and Things*, 93–106, where he considers especially D. M. MacKay's paper on 'Complementarity' in *Aristotelian Society Sup. Vol.* (xxxii), 105–22. In the same volume xxxii, 75–104, P. K. Feyerabend attacks the usual positivist view of complementarity as one which leads to stagnation in science. D. Bohm's consideration of bacterium and spore as complementary forms of the same entity *Quantum Theory*, 162, is less unhappy, but does not resemble our view, which is derived from B. Lonergan, *Insight*, chapter four.

'phenomenological' to 'statistical' which occurs here is misleading.[36] On examination classical thermodynamics can be seen to be no more phenomenological than statistical thermodynamics. It involves an abstract system which provides implicit definition for the terms T, S, etc., and it is related to phenomena through implicit rules of correspondence.[37] So, for example, heat is implicitly defined by the first law of thermodynamics and a correspondence rule is implicit in the origin of, and in the process of verification of, that law. Again, as with all classical laws, the verification consists in a convergence of experimental results on that law.

Classical thermodynamics, in so far as it has nothing to say regarding microstructure, is comparable to classical hydrodynamics with its homogeneous fluids and uniform pressures. The transition from classical to statistical thermodynamics is, as Feyerabend puts it, a matter of 'replacement rather than incorporation.'[38] That replacement is called for by the existence of residual phenomena. The replacement is statistical precisely because the residue is fundamentally statistical: what is conceived classically as a uniform ideal volume reveals itself empirically to be an aggregate, coincidental from more than one point of view even in the case of the 'ideal' gas, thus resisting classical systematization and calling for statistical understanding. But this statistical handling of the problem requires classical definitions, and these definitions are, as usual, definitions which belong to members of the aggregate. This analysis bears an interesting resemblance to our discussion of the Correspondence Principle, and it adds significance to Feyerabend's remark, 'a purely kinetic account of the phenomena of heat does not yet seem to exist. What exists is a curious mixture of phenomenological and statistical elements, and it is this mixture which has received the name "statistical thermodynamics."'[39] Inevitably there are parallel definitions in old and new theory: more especially since the relation of old to new is basically one of mean values to distributions. But this relation was not known or formulated prior to the emergence of the new theory and so the new theory gave new meaning to the old terms T, P, S, etc. So, for example, where pressure meant 'force'—force being defined implicitly by the

---

[36] Joos, for example, divides his treatment into phenomenological and statistical parts: *Theoretical Physics*, London, Blackie & Son, 1951, Parts 5, 6.

[37] Explicit and clearly formulated rules of correspondence are an unattainable ideal. 'The haziness that surrounded such correspondence rules is inevitable, since experimental ideas do not have sharp contours that theoretical notions have.' E. Nagel, *The Structure of Science*, 100. Correspondence Rules for his own particular classical formulation of Thermodynamics are discussed by Giles, *Mathematical Foundations of Thermodynamics*, New York, Pergamon Press, 1964, ch. 1.

[38] P. K. Feyerabend, 'Explanation, Reduction and Empiricism' *Minnesota Studies*, III, 78.

[39] Ibid.

classical equations—now it means 'average force': in both cases there is a classical correlation involved, but in the second case there is also statistical understanding, and the meaning of 'force' has in fact been extended. Also there is a shift in predication: 'pressure' is predicated of the gas as a whole, the 'average momentum' is not.

To go further here would carry us into a. criticism of reductionism which we prefer to postpone to chapter nine. But we may remark that while we largely grant the form of 'reduction' of classical thermodynamics to statistical thermodynamics that Nagel describes,[40] we would not concede that this example is 'a basis for a generalized science.'[41] Pressure may well be accounted for as an average of micropressures, but there are, for example, no 'micro-sights' to which to reduce sight: nor, if there were, would this be considered a proper reduction. The thermodynamics example, indeed, is a peculiar one of 'reduction' to a theory richer in meaning—not at all the normal tendency of the reductionist enterprise. Classical thermodynamics rests, so to speak, on the rather harmless systematization provided by correlations between averages and such like: the significant reduction-resisting systematizations, however, are those which differentiate the sciences, those which order in a higher science what are only coincidental aggregates in a lower science. With these we will deal later.

The account given so far of classical and statistical procedures and explanations has stressed the way in which the two approaches complement one another. That account differs in more than one respect from Hempel's 'Deductive-Nomological vs. Statistical Explanation.'[42] There is of course a basic difference of concern: our interest was in patterns of discovery of both types of law, his was in deductive formulation of the two types of explanation. But this basic difference of concern need not have been a cause of disagreement: nor more than that the difference between Aristotle's Posterior and Prior Analytics should give rise to an opposition between them. Hempel remarks, prior to his examination of statistical explanation, that 'this examination will lead to the conclusion that the logic of statistical systematization differs fundamentally from that of deductive-nomological systematization. One striking symptom of the difference is what will be called here the ambiguity of statistical systematization.'[43] This ambiguity of statistical systematization occupies a good deal of his attention. For him it was a symptom of difference: for us, as we have seen in chapter four, it is a symptom of something else, a misconception of statistical explanation.

To a correct account of the concrete logic of discovery there should be a corresponding account of the logic of prediction and explanation. We will

---

[40] Nagel, *The Structure of Science*, 344–5.

[41] Ibid., 342.

[42] *Minnesota Studies*, Minneapolis, 1962, III, 98–169.

[43] Ibid., 124.

indicate that account firstly in a manner continuous both with our discussion so far of the logic of discovery and with the Aristotelian version of Mill's *A System of Logic*, the Analytics. Later we will express our conclusions in a form similar to that of Hempel's scheme,[44] thus facilitating the comparison of these conclusions with the view of Hempel and others on the structure of explanation. But the discussion of the logic of discovery here is not merely for the sake of this comparison: it is continuous with the general orientation of the work towards the establishment of a science which can be identified with the logic of discovery, an identification which will be made explicit in chapter twelve.

The linkup with Aristotle can be made through chapters one and two of the 2nd book of the Posterior Analytics. In chapter one Aristotle specifies types of inquiry by the type of question one seeks to answer, and he shows in chapter two how these various types of inquiry are concerned with the finding of a middle term of a syllogism. We are interested here mainly in the first two of the four Aristotelian questions: (1) the *that*-question, e.g., that the Sun is eclipsed and (2) the *why*-question, why is it eclipsed. Before going on it is as well to point out the central deficiency of the Aristotelian position. It is the absence of consideration of the existential question which should follow the answer to question (2) or question (4). For Aristotle, the existential question was answered first in establishing the existence of the phenomenon to the explained: the hypothetical nature of the consequent explanation, the answer to question (2) or question (4), was not explicitly appreciated by him. Aquinas was to remedy that defect: '. . . in astrologia ponitur ratio excentricorum et epicyclorum ex hoc quod, hac positione facta, possunt salvari apparentia sensibilia circa motus coelestes; non tamen ratio haec est sufficienter probans, quia etiam forte alia positione facta salvari possent.'[45] Our view of the Aristotelian explanatory syllogism, of its emergence through conjecture and its survival pending refutation, is influenced by the supplementation of Aquinas.

Question (2), according to Aristotle, is a search for a middle term, in some sense the *cause,* and the finding of that middle term requires 'quick wit':

> Quick wit is a power of hitting upon the middle term instantaneously; e.g. if one sees that the moon always has its bright side towards the sun, and quickly grasps the reason, viz., that it receives its light from the sun. Let A stand for "bright side turned towards the sun," B for "receiving light from the sun," C for "moon." Then B (receiving light from the sun) is predicable of C (the moon) and A (having bright side

---

[44] Ibid., 100.

[45] *Sum. Theol.*, Ia, q. 32, a. 1, ad 2m. ['. . . in astrology the theory of eccentrics and epicycles is considered as established, because thereby the sensible appearances of the heavenly movements can be explained; not, however, as if this proof were sufficient, forasmuch as some other theory might explain them.']

towards source of its light) is predicable of B. So A is predicable of C through B.[46]

Since geometry normally provides paradigm cases for the discussion of deductive logic, we may well take a first paradigm case for the logic of discovery from there also. Consider the following extremely simple geometric problem: In a circle, of radius one unit, two diameters are drawn at right angles. From an arbitrary point P on the circumference, two perpendiculars, PA and PB are drawn to the two diameters. The question is, What is the length of the line AB? One must obviously seek some relation of the length, AB, to the radius of one unit. In so far as one is 'quick witted,' has the habit of geometry, the solution is clear. Otherwise one must muse and perhaps draw more lines: and the right line to draw is of course the line joining P to the centre. It may be noted also that the question in its 'why' formulation is, Why is AB equal to the radius?

The most obvious principle of the logic of discovery which the example illustrates is the principle of the relevance of diagram: this principle is Aristotelian;[47] it was given new emphasis by Aquinas;[48] it is implicit in the verification principle of meaningfulness,[49] and in successful pedagogical principles.[50] The second point to note is the general one that principles of the logic of discovery have to do with a stage prior to formulation: formulation is consequent to discovery—in our example the result can easily be cast into syllogistic form, and then it is an object of deductive logic.

The simple example illustrates discovery only in relation to one form of Aristotelian syllogism, but the process is characteristic not only of all forms of Aristotelian argumentation but of all forms of contemporary deduction. One might take a first step towards showing this by exposing the central place of the form of inference, which dominates modern logic, in Aristotelian logic.[51] Again

---

[46] *Posterior Analytics*, Bk. I, ch. 34.

[47] Cf., e.g., *De Anima* III, 7.

[48] Cf., e.g., *Sum. Theol.*, Ia q. 84 a. 7.

[49] Cf. A. Pap, *Introduction to the Philosophy of Science*, 14.

[50] Cf. The works of W. Sawyer: *Mathematicians' Delight*, London, Penguin Books, 1943; *A Prelude to Mathematics*, London, Dover, 1955; *A Concrete Approach to Abstract Algebra*, San Francisco, Freeman, 1959; etc., and the remarks on that work in chapter 8 below.

[51] This is to some extent carried through in Lonergan's 'The Form of Inference,' *Thought*, 1943, 277–92 (republished in *Collection*, London and New York, 1969; vol. 4, Collected Works of Bernard Lonergan, Frederick E. Crowe and Robert M. Doran (eds.), Toronto, University of Toronto Press, 1998, 3–16.) W. Kneale and M. Kneale, *Development of Logic* (Oxford, Clarendon, 1962), 96–100, point to Aristotle's neglect of the form of inference. This neglect leads to an exaggeration of the opposition of the Aristotelian position and modern logic, whose axiom systems are basically formalizations of the properties of inference (cf. for example, Mendelson's *Introduction to*

one may note that even in the strict and more extended derivations of conclusions in modern deductive systems, with fully formulated axioms and rules of inference, the Aristotelian 'quick wit' or what Suppes calls 'intuition' or 'insight' is relevant: '. . . to derive from the axioms what is known as the *right hand cancellation law* ... the crucial step is realizing what substitution in axiom (1) is appropriate ... For with this insight goes the perception that z O z^{-1} = c by virtue of Axiom (3),'[52] and Suppes discusses the problems of strategies to aid in such discoveries.[53] The complete Hilbert programme of formalization requires what might be called the crystallization of such intuitions or insights into axioms and rules of inference. A classic example of such crystallization is that of the casual insight which occurs regularly in Euclid, the insight e.g., that a line which contains a point of one side of a triangle must contain a point of one of the other sides. This type of insight was formulated by Pasch (1890) into an axiom of order. The effect of such crystallization is to liberate us to some extent from diagram. More generally such crystallization shifts our dependence on 'intuition' or diagram to symbols: and when such symbolism is apt, it can contribute significantly to progress. This was the case with Newton's symbolism for derivatives as compared to Leibniz's. On the other hand one would be at a loss to get the square root of MDCDXXXVI, and a basic flaw in Woodger's efforts to axiomatize biology lay in his failure to find an apt symbolism.[54] Within statistics one may note the aptitude of the range zero to one for measuring probabilities, or the significance of matrix symbolism for the theory of stochastic processes.

The limitations of Hilbert's programme are associated with the non-crystallizability of the full range of relevant insights, and the central thesis which expresses this, Gödel's theorem, is an instance of that peculiar type of insight which we have discussed already, the inverse insight. Again, according to Church's thesis, 'recursive undecidability is equivalent to effective undecidability, i.e. nonexistence of a mechanical decision procedure for theoremhood. The nonexistence of such a mechanical procedure means that ingenuity is required for determining whether any given wf is a theorem.'[55]

Mathematical Logic, the axiom system for propositional calculus, 31, and for first-order theories, 57).

[52] Suppes, *Introduction to Logic*, 106.

[53] Ibid., 136–7.

[54] Cf. J. H. Woodger, *The Axiomatic Method in Biology*, Cambridge University Press, 1937: to judge the value of the symbolism one might best consider such an application of it as that cited in footnote 18 above: do the results warrant the complexity?

[55] Mendelson, *Introduction to Mathematical Logic*, 151. There seems to be an interesting connection too between the debated principle of the excluded middle and the replaceability of a middle term by a better one (to be discussed below), but this would take us too far afield.

All this is relevant to the appreciation of the dependence of theory formation and application—statistical or classical—on 'ingenuity' or casual insights. These topics are normally passed over in discussions of theory formation, but they are in fact the central concern of the logic of discovery. That logic is to be formulated, not in axioms, but in canons of procedure which may lead to such axioms. Both axioms and canons are concerned with the determination of the meaning of terms: but where the axioms express that meaning, the canons in a certain sense only expect it, anticipate it. Or, to return to our example of the simple syllogism, one may say that the complete syllogism consists of two 'axioms' and a conclusion through a rule of inference. Such would be the object of discussion in deductive logic: but the logic of discovery concerns the anticipation of the middle term.

Let us pass from geometry to empirical science. We take as an example of an answer to the first question of Aristotle, the 'that' question, the statement that 'the earth moves in an ellipse (with the sun at its focus).' Now there is considerable discussion as to what is or is not a law in physics, how laws are related to theories, whether theories differ from laws or hierarchically from each other.[56] A first step towards clarifying these problems is to note that, strictly speaking, the above statement is mathematical: it is certainly about the earth and its path, but it defines no physical property. It is a necessary prerequisite of such definition indeed, and comes—as Kepler knew too well—at the end of a long investigation, an investigation governed by the canons discussed earlier in this section. It is in the answer to the second Aristotelian question, the why-question, that we find the physics proper: that physics provides implicit definition of mass and gravitational force. In terms of the simple syllogism, this why-question asks for a middle term, and the discovery of Newton provides it. The resulting syllogism is (with obvious omission of sun, etc.):

The earth OBEYS NEWTON'S LAWS,

WHAT OBEYS NEWTON'S LAWS   moves in an ellipse,

∴ The earth   moves in an ellipse.

Various points should be noted regarding this syllogism. The usual major premise is placed second, but this is only a device to help relate the formulation to the process of discovery. First of all, then, the second premise implies the process of integration which yields the general conic and further the general ellipse under certain conditions of velocity of escape. Secondly, prediction in the strict sense of giving future times and locations is absent: but there is implied the general prediction: all things being equal the earth will continue to move in an ellipse. Thirdly, particular predictions can be included by including initial

---

[56] Cf. for example, E. Nagel, *The Structure of Science*, ch. 5.

conditions somewhat as follows: ... what obeys Newton's Law with initial conditions α_i, moves in this (such-and-such a particular) ellipse, ...

Fourthly the emergence of the middle term is associated with the conception of some 'thing,' some locus of predication. Here it is clearly the earth and the sun, whose masses are implicitly defined by the Newtonian law. In other cases it is not so obvious, e.g., in the case of the that-answer, that 'the tracks in the cloud chamber are such-and-such' we could have:

> The tracks in the cloud chamber ARE FORMED BY PARTICLES OF MASSES m_1, CHARGES e_1 ... ,
>
> WHAT ARE FORMED ... thus ... are such-and-such,
>
> ∴ The tracks in the cloud chamber are such-and-such.

Here one may initially be thinking of the tracks as things, or better of the particles as things: one may note here that one may think of an object as a 'thing' without it being a *thing*.

Fifthly, we can give significant meaning to the Aristotelian term *form*—a meaning not entirely foreign to that of Francis Bacon.[57] The term 'form' is merely a name for the to-be-discovered answer to the why-question of Aristotle;[58] it ceases then to be the mystery which many Thomists have made of it; it need no longer be a 'feature' of things which metaphysicians alone can intuit. In our sense of form, Newton's law gives the form, the why, of the motion. Again, in the cloud chamber example we are looking for the *form* of the tracks, not just the mathematical form which was the answer to the that-question, but the physical form: and the answer was in terms of particle properties and forces. Inversely, the tracks are the matter, in an Aristotelian sense,[59] of the question: matter, in this sense, and form are clearly correlative. This brings us, sixthly, to the problem of Rules of Correspondence, the principles of which are so difficult to formulate.[60] These principles must typify and limit the admissible forms, explanations, and in particular cases should bridge the gap between the extreme terms and the middle term which contains the complex of implicitly defined theoretical terms. Thus

---

[57] Cf. the texts and discussion in *Theories of Scientific Method*, ed. E. Madden, Seattle, University of Washington Press, 1960, 65–69.

[58] Aristotle did in fact relate the why-question to all four of his causes (cf. *Physics* II, 7; *Posterior Analytics* II, 2) but the relevant cause here is the formal cause.

[59] We are reminded here of Whewell: 'ideas are not *transformed* but *informed* sensations ... our sensations, from their first reception, have their form not *changed* but *given* by our Ideas,' *History of Scientific Ideas*, London, 1858,1, 40, whose Kantian-Scotist background we of course reject. For a discussion of Aristotle's view, and finer distinctions omitted here, cf. B. Lonergan *Verbum: Word and Idea in Aquinas*; CWL 2.

[60] Cf. Achinstein, 'Theoretical terms and partial interpretation,' *British Journal for the Philosophy of Science* (XIV), 1963–4, 89–105.

masses etc., are implicitly defined by the law of gravitation, but it is the earth and the sun which have masses, and it is the measured ellipse which leads to the law and is a source of its verification. Finally, the hypothesis contained in the middle term is restricted by the verifiability principle of meaningfulness, and this principle corresponds to the Aristotelian view that the form is not a Platonic idea but the form of this phenomenon, the form of this matter. To this latter point we will return both in chapter nine and in the concluding chapter.

Seventhly, the significance of the middle term goes beyond the particular phenomena which gave rise to it: that of our first example has reference not only to planets but to apples, projectiles and comets, and thus it becomes a principle of wider prediction and explanation.

Eighthly, the middle term makes possible its own replacement but, so to speak, resists it. It resists it, because it is the basis of the formulation of further questions and experiments: it keeps theorizing within its own bounds—as in quantum theory at present—and it can be replaced not by logic but by some leap of insight to a new theoretical position with its complex of unexpected terms and correlations. Still, it does make that replacement possible by the method already discussed, the method of residues. Through failure in detail, wider application, technical advances in exposing data, etc., the extreme terms change and the middle term becomes, e.g., not Newton's, but Einstein's Law, and a new syllogism emerges:[61]

> The earth OBEYS the Einstein eqt
>
> $$\frac{d^2x\alpha}{ds^2} + [\mu\gamma, \alpha]\, \frac{dx\mu}{ds}\, \frac{dx\gamma}{ds} = 0$$
>
> what OBEYS ... , moves in an ellipse with perihelion advance Φ,
>
> ∴ The earth moves in an ellipse with perihelion advance Φ.

The discrepancy here between predicted advance and observed advance calls for further investigation, and so on in a type of asymptotic process of improving explanations.

Ninthly, there is a relation between middle term and extreme terms which marks a shift from description to explanation, where these two are here given a technical sense: description characterizing things in relation to us, explanation correlating things among themselves. There is an echo here of Bacon: 'the Form or true definition of heat, heat that is in relation to the universe, not simply in relation to man.'[62] Description and explanation in the above sense, however, are not paralleled by the distinction between secondary and primary qualities. Both

---

[61] Cf. Eddington, *Mathematical Theory of Relativity*, Cambridge University Press, 1924, 85–90.

[62] Cf. *Theories of Scientific Method*, 66 ff, for this and related texts.

secondary and primary qualities can be described, both require explanation: the explanation of extension, for example, being in terms of a geometry of space-time.

Tenthly we may note that the relation between middle term and extreme terms is one of possibility, not of necessity. We are reminded of Peirce's 'abduction' which 'merely suggests that something *may be*,[63] and of R. Harré's underlying condition on any explanatory mechanism, 'we may be entirely wrong'!

That these considerations are not restricted to physics is shown by the following example. The example, in the form of an Aristotelian why-question, is, 'why does the rabbit run to a burrow when the dog comes?' The question, in present terminology, asks for the form of the locomotion. It is in fact *form*[64] of escape reaction, one form of animal locomotion which is of wide comparative significance, for the 'term "escape reaction" is applied not only to "heading for the rocks" (a fish reaction) but to a great variety of responses characteristic of very different species as well.'[65] Full explanation of the escape reaction is of course not reached until the shift from description to explanation is made on all levels involved. But for the sake of illustration we give a simple syllogism, whose middle term is odd but sufficient, covering a cluster of further problems:

> The rabbit SENSITIVELY DIS-APPRECIATES THE DOG
> WHAT SENSITIVELY . . .   runs to a burrow,
> ∴ The rabbit   runs to a burrow.

Switching now from classical to statistical questions we come to consider a type of middle term which Aristotle did not allow for.[66] Nor can one easily arrive at the middle term we have in mind without giving a peculiar twist to Aristotle's why-question. Aristotle was all the more handicapped by the prior that-answer not being adverted to: the that-answer we have in mind is one such as: 'when 10 Prussian Corps were exposed to mule kicks over a period of twenty years, the following frequency distribution of deaths per corps per year was obtained:

Deaths per corps per year:	0	1	2	3	4	Total
Frequency:	109	65	22	3	1	200'[67]

---

[63] *Collected Papers*, Cambridge, Mass., 1931–8, Vol. V, parag. 171.

[64] The discussion of concrete recurrence-schemes in chapter ten will reveal that we are using the word 'form' somewhat loosely here.

[65] M. Beckner, *The Biological Way of Thought*, New York, Columbia University Press, 1959, 122; one might make an interesting comparison with Aristotle's *De Motu Animalium*, 7, 8.

[66] Cf. *Physics* ii, 4–6; *Posterior Analytics*, i, 30.

[67] K. Mather, *Statistical Analysis in Biology*, 36.

To ask 'why?' here is to ask a variety of different questions.[68] In its general Aristotelian sense it would require an answer in terms of the causes of each particular death. The sense which interests us here however is again a sense which looks to a *form*-answer: the sense which asks for the form of the distribution. The discovery of the form of the distribution is the discovery of the middle term, and the result can be cast into syllogistic form:

The deaths OBEY POISSON'S LAW (with $\mu = \sigma^2 =$ such-and-such),

WHAT OBEY . . .   are distributed thus (as in list),

∴ the deaths are distributed thus (as in list).

Just as with the example of the earth's motion, so here various points might be noted, such as the difference between the derived theoretical distribution of the second premise and the sample distribution of the conclusion. However, in view of the parallel between the two syllogisms, and of the earlier discussion of statistical residues, etc., this would be repetitious. But we may note how the syllogistic formulation of statistical investigation would seem to indicate that, if probability is a property of anything it is a property of experimental results. Before continuing the discussion we add two further examples of statistical hypotheses:

1. The momenta of gas molecules ... OBEY the MAXWELL-BOLTZMANN LAW (with particular parameter values),

   WHAT OBEY the MAXWELL-BOLTZMANN LAW ... are distributed so-and-so

   ∴ The momenta of gas molecules are distributed so-and-so.

2. The energy values and distributions OBEY A (particular) SCHRÖDINGER EQT.,

   WHAT OBEY THIS SCHRÖDINGER EQT., are distributed ... 'thus' (E_i and ω_i)

   ∴ The Energy values and distributions are ... 'thus.'

We now come to consider our account of classical and statistical investigations in a symbolic form which parallels that of Hempel:[69] '. . . the explanation is a deductive argument of this form:

$$L_1, L_2 \ldots \ldots L_r$$
$$C_1, C_2 \ldots \ldots \ldots C_k$$

$$E$$

---

[68] Cf. footnote 58 above, 75.
[69] *Minnesota Studies* III, 100.

Here $L_1, L_2 \ldots L_r$ are general laws and $C_1, C_2 \ldots C_k$ are statements of particular occurrences, facts, or events; jointly, these premises form the explanans. The conclusion E is the explanandum statement.'

Our account does not fit easily into this description. In so far as we would try to fit it into the Hempel description, we would rather consider the major premise in our case to contain the complex of laws suitably selected, combined and systematized as described in chapter three, with the inclusion of correspondence rules; the minor premise to contain particular secondary determinations of the laws, initial conditions, particular mass values, etc.; the conclusion to be the union of these as giving an account of the actual situation, or the one being predicted. If the laws in the major premise are of the classical type, we have something parallel to Hempel's Deductive-Nomological version of the symbolic pattern. Our minor premise is then combined with the major premise in the standard fashion of using boundary values, to give the conclusion. But if we consider the laws in the major premise to be of the statistical type then we have something which finds no parallel in Hempel. For these statistical laws are directly explanatory of the minor premise of conditions or secondary determinations, in this way making possible the prediction of the conclusion. To give a simple illustration, the major premise might be the general Normal law, the minor premise a list of heights of people in a sample; a combination of these leads to a determination of the parameters μ, σ, and thus to a particular Normal law for height distribution in a given population.

Thus the two types of explanation are distinguished by the way in which the major and minor premises combine to yield the conclusion. In classical explanation the minor premise is extrinsic to the formulated law: it gives the primary relativity expressed in that law its secondary determinations. In statistical explanation the minor premise is what is systematized by the law. This harks back to the earlier discussion of the way in which the two types of explanation complement one another. To look still further back we recall the discussion of the existence in the general case, of a diverging series of conditions in chapter three. This diverging series falls within the scope of the minor premise. Even *per impossibilem* all these conditions were listed in the minor premise, the deduction of the process—or, most generally of world process—from a major premise of all relevant laws would be at best no more than a piecemeal step-by-step affair: for, the conditions form a spatiotemporal coincidental aggregate. And even statistical science can only go a certain distance towards general knowledge of them.

An obvious question here regards the occurrence of both types of law in the major premise. Such a mixing is in fact possible, and fits into the pattern of our own syllogism of scientific explanation, pointing to the general validity we would claim for it. In chapter eleven, indeed, we will move forward to see how classical and statistical laws can be combined to give some general yet concrete account of world process or evolution. Before doing that we wish to note the relevance of the recurring distinction of primary relation—e.g., that contained in the law $F = m_1m_2/r^2$ and secondary determinations—e.g., numerical values for m_1, m_2, r, in a given case—to the problem of classifying laws of nature.

Consider for example the statement 'lead melts at 327C°.' What sort of law is this? From our point of view, the primary correlations involved are the implicitly defined—by correlation with pressure and volume variations—theoretical states of matter, gaseous, liquid, solid, as well as the definition of temperature—defined implicitly e.g., by correlation with mercury volumes and water states. The law gives the transition temperature for solid-to-liquid state transition of lead: a secondary determination of the general law. But it is to be noted that the statement leads to a further why-question, and to explanation on the atomic, etc., level. Again Balmer's law, as Nagel notes, was not abandoned when the Bohr theory of the atom which explained the law was replaced by quantum theory. 'Such facts indicate that an experimental law has, so to speak, a life of its own, not contingent on the continued life of any particular theory that may explain the law.' [70] Why? Because strictly speaking Balmer's law is a correlation of numbers, not a physical law: like the ellipse of the earth's orbit, it is the object of a why-question. It is a series of numbers, of secondary deter-minations, the explanation of which turns out to be to some extent non-classical. Our distinction and discussion leads also to some fruitful suggestions regarding Braithwaite's hierarchy of hypotheses:

'Let us consider as an example a fairly simple deductive system with hypotheses on three levels ...

The system has one highest level hypothesis:

I.   Every body near the earth freely falling towards the earth falls with an acceleration of 32 feet per second per second.

From this there follows, by simple principles of the integral calculus, the hypothesis:

II.   Every body starting from rest and freely falling towards the earth falls $16t^2$ feet in t seconds, whatever number t may be.

From II there follows the infinite set of hypotheses:

IIIa. Everybody starting from rest and freely falling for 1 second towards the earth falls a distance of 16 feet, etc.'[71]

From our point of view the highest level hypothesis is an approximative particularization to a class (m_i, M_e) of the general law defining gravitational

---

[70] Nagel, *The Structure of Science*, 86.

[71] Braithwaite, *Scientific Explanation*, Cambridge University Press, 1953, 12.

masses: moreover, it owes its 'height' not to anything rarefied about the terms defined, but to the abstractness of a second order differential equation.[72] The first integral, which is omitted, comes nearest to giving the concrete *form*[73] of the fall with secondary determinations of initial position and velocity, and of course a determined value of m_i. The second level hypothesis is a type of 'freezing' of the phenomenon through further integration, but again has nothing to do with the 'abstractness' of the theoretical terms involved. The third level hypotheses are merely the result of a selection of secondary determinations of t, whose definition is assumed throughout.

In chapters ten and eleven we will face the task of providing a mixed—classical and statistical—middle term of a scientific syllogism, which will be in some way an explanation of world process, of which evolution is a part. The extreme terms must obviously provide a description of that process, a description which will include some classification of events, with their number and distribution. Such a description may be considered as a 'that-answer.' The consequent why-question cannot be of the straight classical type, since we are interested in numbers and distributions, nor can it be of the ordinary statistical type, since a very evident part of the description would be the established temporal series of related types and organisms. The question 'why is world process such-and-such?' will, then, have characteristics of both types of explanation discussed, and it will have the characteristics common to both which we already listed.[74] It asks for a hypothesis which can be filled out, modified, verified, improved on asymptotically; for the *form* of world process.

---

[72] Cf. Lindsay and Margenau, *Foundations of Physics*, 29–48.
[73] Cf. footnote 64 above, p. 77.
[74] Cf. above, pp. 74–77.

Randomness, Statistics, and Emergence

In his *Probability and Induction* W. Kneale levels a variety of criticisms against the frequency theory of probability, especially as defended by von Mises. We will be concerned here mainly with two of these criticisms. Firstly, Kneale remarks that 'the theory of von Mises does not elucidate the conception of chance, but leads on the contrary to a very strange confusion between chance and law.'[1] That von Mises has failed to give any account of chance is certainly a Legitimate criticism: but we would hope to show that this account can be given within the frequency theory, that the frequency theory does not inevitably lead to a confusion of chance with law,[2] that in fact it makes room for an essential advance on previous analyses of chance. Kneale's other objection is that the frequency theorist rejects any idea of looking further for an explanation, for example in terms of the physiology of reproduction, of why the probability of male babies is .52. This is certainly not true, as our discussion in chapter five of the actual scientific use of the frequency theory has shown. Statistics indeed lead us to look for such further explanations, at times when we would not otherwise have suspected the need for further investigation. This was most clearly illustrated by our analysis of the *Ranunculus* experiments. Kneale's conclusion to his objection had best be quoted:

> All this (search for explanation) the frequency theorist rejects, at least by implication. To the question "why should we find order on a large scale combined with disorder on a small scale?" he replies in effect "Because the larger the scale the more the order." If his assertion makes sense at all, it explains what puzzles us only in that Pickwickian sense of "explain" in which the statement that there are two lions in my garden explains why there is one. He has not, as is sometimes said, repudiated the view that there are strictly universal laws of nature. On the contrary, he is committed to maintaining that there are such laws. But he has to say that they are *laws of chance,* and that the most funda-mental among them are concerned with entities which are certainly not fundamental, namely, infinite sequences of fractions of relative frequency. This is surely a mistake.[3]

Kneale's two objections are obviously closely related, and, just as he was correct in pointing to von Mises' failure to discuss chance, so he is correct here in accusing the frequency theorists of not giving a sufficient answer to the above question about large scale order. There are of course answers which seem better than the one which Kneale attributes to the frequency theorists: one can, for

---

[1] *Probability and Induction,* 162.

[2] This is the paradox mentioned above, p. 18.

[3] *Probability and Induction,* 164.

example, draw on Poincaré's discussion of the complexity of causes and the influence of small variations etc.[4] and elaborate reasons such as those quoted earlier from Arley and Buch.[5] But in fact these reasons are more concerned with the occurrence of randomness and the need for statistics rather than with the occurrence of order or law at this level. They are then more concerned with von Mises' axiom of randomness than with the axiom of limiting frequencies, and, as Kneale points out, it is the latter axiom which needs justification. The former axiom is somewhat negative, relating more to disorder; the latter axiom touches more the positive foundation for the existence of laws associated with that disorder. There is lacking, however, any explanation of how this can be in the way the frequency theorists claim it to be. We would hope to fill this lacuna in the frequency theory treatment, and this in a manner which will give a single coherent answer to both of Kneale's objections.

It would help to consider firstly Kneale's own position. The most obvious criticism of that position is that his considerations are altogether remote from scientific practice, with such a remoteness indeed as could lead him to the false conclusion that the frequency theorist neglects causal explanation. This deficiency in Kneale, and his overconcentration on clarifying 'The meaning of probability statements made by plain men,'[6] have already been adequately discussed by F. J. Anscombe.[7] Mr. Anscombe's interests, on the other hand, clearly coincide with our own:

> My own interest in the subject is more limited. I should like to see clarified the probability statements made by scientists i.e. I should like to know how scientists use probability concepts in their work, and, if possible, how they *should* use them. Once this has been done, the probability statements of plain men should prove fairly easy to describe, since, when not fallacious, they would presumably be found to be approximations to the procedures of the scientists.[8]

We have, however, one reservation about Mr. Anscombe's view: the procedure of plain men does not necessarily coincide with that of scientists, but their statements need not for that reason be considered 'fallacious.' The justification for this disagreement lies in our discussion of the procedure involved in reasonable betting, and our conclusions there correspond remarkably well with Kneale's conclusions about probability statements. A further justification lies in the fact that there are two distinct notions of probability, and so two distinguishable sets of probability statements that require analysis. As yet,

---

[4] *Science and Method*: English translation, in *Foundations of Science*, 395ff.
[5] Chapter 2, pp. 11–12.
[6] *Probability and Induction*, 158.
[7] 'Mr. Kneale on Probability and Induction,' *Mind* LX (1951), 299–309.
[8] Ibid., 299.

however, we have not discussed the second notion: it will occupy us in chapter seven.

E. Nagel's criticism of Kneale is also relevant here:

> Mr. Kneale believes that empirical frequencies can serve as evidence for estimation of the ratios of ranges. But why should the empirical relative frequency with which A's are B's be relevant for estimating the ratio of the ultimate possibilities subsumed under these terms? ... by what line of reasoning do such empirical frequencies become relevant for estimating ratios of ranges? If Mr. Kneale supplies answers to such questions, they have escaped me. Indeed, I do not think he can answer them satisfactorily, for the theory of ranges has no point of contact with practice and with empirical frequencies, unless it is coupled with assumptions which connect in a *de facto* manner probabilities with relative frequencies.[9]

Nagel's criticism of Kneale parallels in a certain way von Mises' exposure, to be discussed in chapter eight, of Poisson's implicit assumption of the frequency theory in his treatment of the Law of Large Numbers. The situation here is one in which what may be called a casual insight, an assumption not easily detected, requires crystallization in an axiom. We will deal in detail with that type of problem in chapter eight, but we may note here how such implicit assumptions can escape attention, as was the case in Euclid's assumption of the axiom of order, or in the implicit assumption of the axiom of choice in many proofs in mathematics. Such casual, unformulated, insights may run, as it were, uncontrolled through an entire treatise. Kneale's range theory of probability would seem to suffer from such an inadequacy.

C. D. Broad, in a lengthy discussion of Kneale's work,[10] centres his attention on what is perhaps Kneale's most important contribution to the problem of induction, and in doing so he throws new light on Kneale's objection to the frequency theorists:

> The frequentists are quite right in saying that the evidence for such (probability) rules is observed frequencies. Their mistake is to hold that what is inferred is definable in terms of frequency. This mistake is analogous to that of thinking that a *law* is a hundred per cent *de facto* association. The assumption at the back of both mistakes is that the conclusion of an inference must be a proposition of the same type as the premises. If Mr. Kneale is right, the conclusions of all ampliative inductions are different in kind from their premises. For the premises

---

[9] E. Nagle, in a review of Kneale's book, *Journal of Philosophy*, 1950, 548.

[10] *Mind* (LIX) 1950, 94–115.

are in all cases about *matters of fact*: whilst the conclusions, according to him, are *principles of modality,* whether they be laws or probability rules.[11]

Now von Mises would certainly dissociate himself from the mistaken view that Broad describes: nonetheless this view does recur in discussions of the meaning of probability statements and when it does it creates problems such as that of the 'looser fit of probability statements' which so tests the ingenuity of Braithwaite.[12] Kneale's view puts him beyond such difficulties but, as Broad points out,[13] his view certainly requires further clarification.

Kneale would hold that what are reached by intuitive or ampliative induction are *principles of modality* which are to be contrasted with matters of fact.

> It is a fact that my pen is red, but it is a principle that a cylindrical thing can be green or again that a thing which is red cannot also be green at the same time ... Principles are concerned with possibility or impossibility, necessity or nonnecessity, and they are in a sense more fundamental than facts, since they determine what facts there can be.[14]

With this view is closely linked his conclusion regarding the nature of scientific theories:

> It is felt as an imperfection of a theory that it should assume laws which cannot be seen to be intrinsically necessary. This does not mean, however, that we can hope to derive laws of nature some day from self-evident truths alone. Although the connections *within* the world of transcendent entities posited by a theory may be self-evident, the relations *between* this world and the world of perceptual objects remain opaque to the intellect ...[15]

There is a certain echo here of our discussion of laws and abstraction in chapter three, and it is worthwhile to recall certain aspects of that discussion in terms of the work of Hoenen.[16] We may quote M. Whitcomb Hess's review of Hoenen's *Reality and Judgment according to St. Thomas* where he summarizes Hoenen's achievement:

> His thesis of "the Nexus"—which is that the sensible data as given in experience must already contain the necessary structural nexus if they

---

[11] Ibid., 105.

[12] *Scientific Explanation*, chs. 6 and 7. Cf. chapter 7 below.

[13] *Mind* (LIX) 1950, 97–8; 105; 111–115.

[14] *Probability and Induction*, 32.

[15] Ibid., 97.

[16] Like Kneale's, Hoenen's work was carried out before the war and appeared then in a series of articles, those already referred to in chapter three, footnote 24. He later brought his results to fruition in the book mentioned in the text.

are to lead up to the judgment—is one that has been mentioned by very few Thomists and utilized by none since the Thomist renaissance. This nexus, which Aquinas described as "the necessary mutual relationship of terms in a proposition," is found operating in the trenchant caesura which occurs between the apprehension of data and the judgment.[17]

If Lonergan's analysis[18] is correct, Hoenen's view, and to a lesser extent Kneale's, would constitute a move towards a rediscovery of the Aristotelian position, a position obscured by the Scotist view directly inherited in Oxford, indirectly inherited through the nominalists by the scholastics and the pre-Kantian rationalists.

As we shall see, Kneale succeeds no better than von Mises in giving a satisfactory account of chance. The difficulty in giving such an account lies in fact in what Eddington calls the 'irrationality' of chance,[19] the fact that, as Hume remarked, 'the very nature and essence of chance is a negation of causes.'[20] But if this is so, an adequate discussion of chance should certainly expose the reasons for this absence, or apparent absence, of lawfulness, this negation of causes. Spencer-Brown hits on the right approach:

> Let us suppose we have an old car, and that on one occasion when we are unable to start the engine Mr. X is in the car. If our attention were drawn to this conjunction of events we should probably say that they "had nothing to do with one another," that they just "chanced" to be coincidental and that their correlation was "not significant."[21]

Our task is to make precise this absence of significance and the relation of this absence to the conjunction and coincidence of events.

Our discussion here is not one of efficient causes or their absence: it is concerned with the presence or absence of laws, of correlations. In this Kneale would agree: 'we speak of chance when we believe that some characters are connected by law and others not.'[22] It is the same notion of lawful connection and its absence that is in question in Bunge's treatment of the present problem.[23] Also, his view that 'causality involves artificial isolation' parallels our own account of causal laws as being conditioned by the 'equality of other things,' and his discussion of intersecting chains and processes serves to illustrate our remarks on

---

[17] *Philosophical Review*, 1953, 627.

[18] Cf. *Verbum: Word and Idea in Aquinas*, CWL 2.

[19] *Philosophy of Physical Science*, Cambridge University Press, 1949, 180.

[20] *A Treatise of Human Nature*, Oxford, 1896, 128.

[21] *Probability and Scientific Inference*, London, Longmans, 1958, 43.

[22] *Probability and Induction*, 180.

[23] M. Bunge, *Causality*, Cambridge, MA, Harvard University Press, 1959.

the diverging series of conditions on any process. His summary account of chance is worth quoting in full, presenting as it does a clear modern formulation of the Aristotelian position on the question:

> If the causal principle and the theory consisting of its unlimited extrapolation allow for *independent* or parallel causal lines, then they are consistent with the admission of that type of chance consisting in the crossing or encounter of mutually independent causal lines— independent, that is, till the instant of their meeting. Hence although the causal principle excluded chance in connection with every *single* sequence, it does not exclude contingency as synonymous with independence (or mutual irrelevance) of different causal series. This sort of chance, dealt with in antiquity by Chrysippus and in modern times by Cournot, is not a mere name for human ignorance—save as regards the prediction of exact place and date of the chance encounter. Even the exact foreknowledge of such a coincidence would not prevent its being a chance encounter, since up to the moment of their crossing the two lines were disconnected; they are objectively contingent upon each other—to the extent to which independence, or irrelevance in definite respects, has an ontological status.[24]

As we shall see, there is room for improvement in this account of chance, in the specification of its relation to human ignorance and as regards its objective status.

One of the senses of chance, the third, discussed by Nagel[25] coincides with that of Bunge. Nagel remarks on the similarity of his own discussion of 'chance' to Aristotle's analysis of 'accident,' but points to a fundamental disagreement between them,[26] due to what he calls Aristotle's 'absolutist' or 'essentialist' view of definition. Aristotle's position, however, both on 'accident' and 'definition' is perhaps more nuanced than Nagel is led to believe. We have already noted how Kneale's view on laws and their origin had some resemblance to that of Aristotle. A more extended discussion of Aristotle's view is therefore called for in order to expose the basis of the divergence of the other views from his, and also in order to account for the manner in which Kneale develops his rather un-Aristotelian position on chance.

First of all, there is a duality in Aristotle's conception of accident, for accident has two distinct opposites. It is defined against essence, principally in the *Metaphysics*;[27] it is defined against existence, principally, if more implicitly, in the *Categories*.[28] It is the former notion of accident which is most relevant to the

---

[24] Ibid., 99.

[25] *The Structure of Science*, 316–9.

[26] Ibid., 329, footnote, 42.

[27] E.g. *Met.* V. 1025a 14.1.

[28] *Categories* 1.

discussion of science.[29] But the latter notion might well be considered in relation to probability, for accident as defined against substance is related to the question of existence[30] or occurrence in (if it is an accident) or not in (if it is a substance) a subject, and we have already seen, particularly in chapters three and five, how the questions of statistics, 'the laws of chance,' are normally questions not of essence, of 'what,' but of existence and occurrence, of the number of actual existents and occurrences, of actual distributions. However, these are only random indications not directly related to our present task. We make no effort here to adequately interpret Aristotle.[31] Still, the underlying Aristotelian position which leads to a coherent account of these notions may be presented in the excellent summary form of M. Novak. Its relation to the position of Hoenen mentioned above will be seen immediately, but it should also be seen as the background to our discussion in chapter three of abstraction, of laws, and of the non-systematizability of the aggregate of particulars.

> One of Aristotle's basic problems was to work out a notion of science that would account, at the same time, for the flux of individuals and the necessity and universality proper to science. To this end, he seized on insight into phantasm as the central, indispensable, and critical activity of human intelligence: "No one can learn or understand anything in the absence of sense, and (2) when the mind is actively aware of anything it is necessarily aware of it along with an image" (De Anima, III, 8:432a 14–17). In this act the individuality of the concrete particular is present: there, within it, and nowhere else, is grasped the necessity required for science. In a second moment there flows from the insight the inner word or concept or definition. The concept or definition is the *universalization* of the insight. The insight itself is bound up with the image. Thus the first moment of understanding, bound up with imagination, occurs at the crossroads between the particular and the universal. The second moment passes over into the realm of the universal, through conceptualization. Insight grasps a new middle term; conceptualization formulates it; a new definition is ready for science. Insight grasps the unity that constitutes a sensible substance; conceptualization and expression in propositional form make possible a judgment of true and false, an affirmation or negation of existence (Met. VI, 4). This activity which is the starting point for reaching

---

[29] *Met.* V. 1027a 20–21; *An. Post.* I. 75a 13–14, 31–33.

[30] *Met.* V, 7 esp. 1017a19 1–23.

[31] In particular we note that we have consistently avoided the question of substance: we would consider that problem adequately handled elsewhere. Cf. M. Novak, 'A Key to Aristotle's Substance,' *Philosophy and Phenomenological Research* (24), 1963–4, 1–19; B. Lonergan, *Insight*, CWL 3, chs. 8, 15.

essence, science, substance, and existence, is the discriminating factor in Aristotle's metaphysics ... It is the key to Aristotle's escape from Plato's intellectualism.[32]

To complete the picture one must note Aristotle's distinction between primary substance and secondary substance: secondary substance can be identified with the definition reached through conceptualization, but primary substance is a 'this,' reached only in the attention that joins insight to sensible detail. It is of the latter that Aquinas speaks when he discusses indirect knowledge of the singular.[33]

With this background one is in a position to appreciate Aristotle's remark about accidental being, 'that there can be no scientific treatment of it.'[34] For, scientific knowledge has to do with the universal and necessary: moreover, the universality is the result of what we earlier called enriching abstraction,[35] and the necessity is one of conceptual relations, arising from that insight into phantasm. Yet while necessity is involved in the relations between the conceptual elements, the conceptual scheme as a whole is grasped as no more than a possibility,[36] and whether that possibility is realized or not raises the question of existence. In modern terms, the scientific theory contains in itself analytic propositions and implicit definitions which express necessary relations—for example, $PV = C$— but the theory itself is a possibility, and if it is not verified, or if it is falsified, it is nothing more than a possibility.

Aristotle does not, of course, deny knowledge of the singular or of the accidental: in a less nuanced way than Aquinas he would grant knowledge, for example, of the flute player who was also a builder. But he would deny systematic knowledge, or science, or a law explaining 'causally' precisely the occurrence in one thing of these two possibilities. Here the notion of accidental thing is mixed with that of accidental cause: 'Things *do,* in a way, occur by chance, for they occur incidentally and chance is an *incidental cause.* But strictly it is not the *cause*—without qualification—of anything; for instance, a housebuilder is the cause of a house; incidentally, a flute player may be so.'[37] But this mixing involves no confusion: 'oportet enim quod unumquodque, sicut se habet ad esse, ita se habeat ad hoc quod sit causa ... homo albus est unum et ens per accidens ... (et ita) causa per

---

[32] M. Novak, 'Towards Understanding Aristotle's Categories,' *Philosophy and Phenomenological Research* (26), 1965, 119–121.

[33] E.g. *Sum. Theol.* I, q. 86, a.1. 'Utrum Intellectus noster cognoscat singularia.' ['Whether our intellect knows singulars']

[34] *Met.* VI, 1026b 4–5.

[35] Cf. above, p. 35 ff.

[36] Cf. B. Lonergan, 'A Note on Geometric Possibility.' *Modern Schoolman* (27), 1949–50. Republished in *Collection*, London and New York, 1969; CWL 4, 92–107.

[37] *Physics* ii, 5, 197a, 11–14.

accidens.'[38] The chance occurrence results in an *ens per accidens*, which in its turn can give rise to chance. It is not that any of the causal sequences leading to the *ens per accidens* is indeterminate, but that they are not mutually determining: indeed their independence leaves room for an indefinite number of possible *incidental causes*: 'That which is *per se* cause of the effect is determinate, but the incidental cause is indeterminable, for the possible attributes of an individual are innumerable.'[39] Nor is it difficult to pinpoint the source, the possibility, of this indeterminateness: as St Thomas remarks in his commentary on the *Peri Hermeneias*, 'Assignat enim rationem possibilitatis et contingentiae, in his quidem quae sunt a nobis ex eo quod sumus consiliativi, in aliis autem ex eo quod materia es in potentia ad utrumque oppositorum.'[40] With the Aristotelian notion of matter has already been associated the notion of empirical residue,[41] the notion of elements of experience which call for no explanation because there is none to be had. Before going on to discuss the changes required in the general Aristotelian position on chance, especially because of the development of statistical science, it would be well to pause in order to compare and contrast this position with that of Kneale.

To begin with the last point mentioned: there is no allowance in Kneale's position for an empirical residue and so it would seem to follow that while obviously[42] 'being identical with a certain individual is an ultimate alternative,'[43] specifying an ultimate alternative would be the less obvious task of specifying the individual. Kneale's individual would be definitionally, conceptually, distinct, a view which clashes with Aristotle's theory of knowledge of the singular. The same oversight with regard to an empirical residue is revealed in Kneale's claim, 'if all the variable magnitudes in question are distances, the configuration space will be identical with the space of common speech.'[44] One may ask what the definitional distinction will be between different locations in this 'space of common speech.' If it springs from the presence of other things, then it is not just a characterization

---

[38] St Thomas, *Contra Gentiles* III, LXXIV; cf. also In Phys. L. II, 1, 8–9. ['Each thing must stand in the same relation to the fact that it is a cause, as it does to the fact that it is a being . . . white man is a unit and being accidentally ... (and so) a cause accidentally.']

[39] *Physics* ii, 5, 196b, 29.

[40] Liber I, lectio 14. ['He {Aristotle} gives as the reason for the possibility and contingency in the things we do the fact that we deliberate, and in other things the fact that matter is in potency to either of two opposites.']

[41] Above, 35 ff.

[42] F. L. Will, 'Kneale's theory of Probability and Induction,' *Philosophical Review* (63) 1954, criticises this taking of individuals as ultimate alternatives even in the finite case, 26–9. He also points out the seriousness of the indeterminacy of ultimate alternatives which Kneale avoids only pragmatically, 24–5.

[43] *Probability and Induction*, 171.

[44] Ibid., 182.

on the basis of the configuration space. If it springs from the space itself, then it presupposes a geometry or an ordering of that space, and that ordering presupposes something to be ordered which we have called the empirical residue.

Kneale comes closest to Aristotle's position in his treatment of the origin of principles of science and the way in which *necessity* is involved in them.[45] But instead of the clear Aristotelian view on law or theory as mere possibility prior to verification[46] we have Kneale's 'opaqueness' of the relation between intellect and perceived object[47] and a confusion concerning possibility. One is reminded here of Max Black's discussion of *Possibility*:

> Perhaps talk about them (possibilities) is simply an indirect way of talking about the rules and the configuration, both of which can be regarded as "actual"? And perhaps talk about possibilities is merely a roundabout way of talking about facts of observation and the relevant laws of nature? For a determinist, this line of thought can easily seem to entail a denial of "objective possibilities." If the relevant laws of nature were known, it would then be seen that whatever happens must happen; the unknown laws that obtain in nature leave no room for unrealized possibilities, and our talk about them is no more than an expression of partial ignorance. Only in what might be called a myopic view of the universe is the word "possible" needed.[48]

One wonders indeed whether Kneale is dealing with equipossibles or equiactuals, and the example to be discussed below leads one to suspect that the latter is somehow the case.

Kneale is led to this position, despite his Aristotelian tendencies, by what we earlier called the impoverished replica view of abstraction.[49] On that view, every aspect of sensible data will have its impoverished replica. Inversely, from knowledge of the totality of impoverished replicas one could construct the totality of sensible data. Again, if one tends to consider statistical laws as a mere cloak of ignorance then one equivalently claims that all the impoverished replicas are systematically related: for, as we have seen, where there is absence of system, there is a place for statistical laws. Coupled with this claim is the claim that the classical, causal, laws are objective in a concrete way and so the systematic relations must hold between the concrete aspects of sensible data. That of course leaves the concrete universe not just determinate—which we would hold to be a reasonable view—but systematic. All these points find their echo in Kneale's discussion. The

---

[45] Cf. the quotation above.

[46] Aristotle did in fact understress verification, but we speak here more of Aquinas' expansion of Aristotelianism.

[47] Cf. *Probability and Induction*, 97, cited above.

[48] *Journal of Philosophy*, vol. LXVII, 124.

[49] Cf. above, p. 44.

characters of whose combination he treats[50] are suspiciously like impoverished replicas. Again, his view, that if all the laws of nature were known,[51] and that they were all well known, one could produce a set of ultimate alternatives, parallels the abovementioned derivation of the totality of concrete data from full knowledge of all replicas.

Furthermore, the same general 'impoverished replica' view fails to notice the distinction between laws and boundary values, initial conditions etc., and so it would aspire to include the numerical values, the secondary determinations of the relations given by the laws, in the laws themselves. 'The determinist begins by overlooking the fact that a concrete inference from classical laws supposes an insight that mediates between the abstract laws and the concrete situation; and once that oversight occurs there is precluded a discovery of the difference between systematic processes and coincidental aggregates.'[52] This same oversight is present in Kneale's work, and so his characterization of ultimate possibilities descends to such details as 'being 3 inches from the nearest hydrogen atom.'[53] Kneale however does not go as far as to say that all characters are systematically related: 'no intelligent determinist wishes to maintain that any two characters we choose to name are connected by a law' and so 'there is no inconsistency whatsoever between chance in the ordinary sense and any plausible theory of determinism.'[54] By denying lawful relations between some of the characters he makes room for his own version of statistical laws: as well as the mode of necessity governing the relations between the systematically-related characters, there is a mode of possibility between the independent characters, where 'this use of the word "possibility" corresponds to one use of the word "chance."'[55] So, as well as a gravitational law relating masses, there will be laws of the type which some way explain the 52 percent relation between the characters of being a male or a female human baby.

At no stage does Kneale concretely illustrate the procedure in determining a law in such a case: or, say, in the case of the biased die, the relations $\frac{1}{6} + a_i$, $a_i \neq 0$, i =1, 2, 3, 4, 5 or 6, for i being uppermost with this tossed die. Let us grant him knowledge of all laws; we may even grant him the additional information that in the actual universe .52 of all babies born were (are, will be) male, or that $(6a_5+1)/6$ of all tosses of a particular die turn up a 5. How is he to proceed? Even though the two cases involve ratios of finite numbers, he cannot consider explanation unnecessary—he would then fall under his own condemnation of the

---

[50] *Probability and Induction*, 66, 174 ff.

[51] Ibid., 174, 179.

[52] B. Lonergan, *Insight*, 97; CWL 3, 120–21.

[53] *Probability and Induction*, 176.

[54] Ibid., 116.

[55] Ibid., 170.

frequentists.[56] But if he is to explain the ratio number .52, he must include in his configuration space the 'space of common speech,'[57] not however merely as additional non-influential dimensions, but as related to the characteristics at each (x_1, t). He will in fact be ultimately involved in a concrete deduction of world process such as we discussed in chapter three, and for that deduction even knowledge of all the laws of nature is not sufficient: he would also need at least information on the totality conditions.

> There results the peculiar type of impossibility that arises from mutual conditioning. Granted complete information on a totality of events, one could work out from knowledge of all laws the concrete pattern in which the laws related the events in the totality. Again, granted knowledge of the concrete pattern, one could use it as a guide to obtain information on a totality of relevant events. But the proviso of the first statement is the conclusion of the second; the proviso of the second statement is the conclusion of the first; and so both conclusions are merely theoretical possibilities.[58]

Thus, the problem of explaining, for example, the .52 of male babies calls for a deduction of a concrete convergent series of processes for each birth, and unless one assumes an 'all things being equal' isolating condition, the deduction cannot even get under way: for, the space of coincidental events and intersections of causal processes would seem to be topologically as complex as 'the space of common speech.'

Kneale's discussion of the 'space of common speech'[59] and of such details as 'being 3 inches from the nearest hydrogen atom'[60] fits easily into Strawson's discussion of the identification of individuals.[61] Strawson's treatment however lays emphasis on the need for *demonstrative identification*,[62] an identification which, as we shall see more clearly presently, lies outside the realms of formulation, and therefore creates a problem in respect to the 'specification of the ultimate alternatives' of Kneale.

Demonstrative identification can certainly make use of the strategy of frames of reference,[63] but use of a convenient frame of reference cannot be equated with a conceptual or scientific distinction of things. For one thing, as

---

[56] Ibid., 164.

[57] Ibid., 182.

[58] B. Lonergan, *Insight*, 650; CWL 3, 673.

[59] *Probability and Induction*, 182.

[60] Ibid., 176.

[61] *Individuals: An Essay in Descriptive Metaphysics*, London, Methuen, 1959, part 1.

[62] Ibid., 18–19.

[63] Ibid., 22, 25.

Strawson points out,[64] we can conjure up a multiplicity of frames of reference according to need and context, and such a possibility warrants the suspicion that we are not specifying the individuals in anything but an ostensive sense. If indeed we use the word 'specify' in its strict sense, additional information such as '3 inches from the nearest hydrogen atom' would not count as specification at all. But what strict meaning of 'specification' are we using here? We can recall immediately our discussion of a distinction between the primary relativity contained in laws of science and secondary determinations such as are given by boundary conditions. But we wish here to elucidate the point from another angle.

Strawson himself gives the basic clue to the solution of the problem of the distinction between specification and, for example, labelling with arbitrary coordinates when he remarks that the framework required for the identification of particulars 'is not something extraneous to the reality of which we speak.'[65] Following up this clue, however, leads us away from Strawson's discussion and into the realms of modern nonstatistical physics.

We turn aside then to consider the problem of specification in the context of modern geometry and physics. First of all let us consider the problem of specifying a geometry or of answering the general question, 'What is a geometry?,' equivalent to asking 'How would one specify a geometry?' For many mathematicians of course a geometry is what mathematicians find convenient to call a geometry, and each such geometry may be defined in terms of a set of axioms. But it would seem that the definition of a geometry in terms of such axioms is only a virtual definition—in the sense that the theorems of the geometry are only virtually given by those axioms. To the question, 'What is Euclidean geometry?,' the answer in terms of its axioms is not complete. Someone who understands adequately what Euclidean geometry is has gone far beyond the axioms. That more adequate characterization would seem to have been the basic insight of Klein when he defined a geometry as a system of definitions and theorems invariant under a given group of transformations.[66] Klein's Erlanger Programme, which was based on this definition, proved however to be too restricted.[67]

> With the advent of Relativity we became conscious that space need not be looked at only as a "locus in which," but that it may have a structure, a field theory, of its own. This brought to attention precisely those Riemannian geometries about which the Erlanger Programme

---

[64] Ibid., 38.

[65] Ibid., 39.

[66] Cf. for example, F. Klein, *Geometry*, New York, Dover, 1956. In the latter part of the book, esp. 174, 184, Klein compares his approach to the axiomatic approach. I have in fact only touched on that question in the suggestion above regarding 'virtuality.'

[67] Cf. E. T. Bell, *The Development of Mathematics*, New York, McGraw-Hill, 1945, 443–448 and the references given there.

said nothing, namely those whose group is the identity. In such spaces there is essentially only one figure, namely the space structure as a whole. It became clear that in some respects the point of view of Riemann was more fundamental than that of Klein.[68]

But the change in viewpoint retained the key insight of Klein into the significance of invariance as a principle of specification. The switch of attention was to differential invariants.

A differential invariant is an abstract object which has in each coordinate system a unique set of *components,* each component being a function of the coordinates and their differentials. For example a quadratic differential form is an invariant which has a single component in each coordinate system, this component being a function which is a homogeneous polynomial of degree two in the differentials and an analytic function of the coordinates. The theory of one or more such invariants is what we call a geometry.[69]

The possibility of classification along these lines is, as Veblen remarks elsewhere,[70] an inevitable generalization in the spirit of the Erlanger Programme.

In the familiar example of Riemannian geometry, the basic invariant is the interval defined by $ds^2 = g_{ik}dx_idx_k$, where the g_{ik} are functions of the coordinates. Now, instead of using different sets of transformations as a source of distinction, the different manifolds are distinguished by restrictions on the coefficients g_{ik}. A basic restriction on the coefficients is that they be components of a covariant tensor of the second rank, for tensors are defined precisely by their transformation properties and the requirement of invariance of $g_{ik}dx^idx^k$ is automatically satisfied. Further restriction to particular covariant tensors gives the various distinct Riemannian manifolds. All this has become familiar through the work of relativity physicists of the past sixty years. So, for example, there is the general distinction between the symmetrical coefficients of gravitational theory and the skew-symmetrical coefficients called in to make possible an account of electromagnetic phenomena,[71] or the particular classification of tensors in more recent work.[72]

---

[68] O. Veblen and J. H. C. Whitehead, *Foundations of Differential Geometry*, Cambridge University Press, 1932; quoted in Bell, op. cit., 443.

[69] O. Veblen, 'Differential Invariants and Geometry,' *Atti del Congresso Internaz. dei Matematici*, 1928, vi. Tomo 1, 183.

[70] 'Invariants of Quadratic Differential Forms,' *Cambridge Tracts in Mathematics* (24), 1927, 49.

[71] Cf. E. Schrödinger, *Space-Time Structure*, Cambridge University Press, 1950, ch. 12.

[72] Cf. for example the three papers by E. Newman, J. N. Goldberg and R. P. Kerr in the *Journal of Mathematical Physics* (2) 1961, dealing with the Petrov classification of Einstein Spaces. Another approach to classification is that of Carl H. Brans, *Journal of*

In thus turning our attention from geometry to physics we return to the clue noted in the quotation from Strawson. That clue has become a commonplace of contemporary physics, and it has its origin in Riemann's *Habilitationsschrift* of 1854. Contemporary physics certainly leaves no room for space as a container, but it goes a great deal further than this. 'Physical space is no longer a locus in which objects are moved about, but Space-Time is itself the only object studied in a complete geometry. There is no such thing as a body in space, but matter is an aspect of the space-time structure.'[73] The latter phrase, however, betrays an unhappy stress, and the balance can be restored by inversion: space-time structure, rather, is an aspect of matter. More accurately, the properties of physical objects find their explanatory expression through a geometry of space-time, and as physics continues to uncover new complexities of relations between physical objects, that geometry can be expected to involve a range of tensors expressing different sets of laws.[74] What that geometry is, is obviously an empirical question. A central problem of its nature is that raised by attempts to evolve a unified field theory.

> As Einstein has pointed out, these are two distinct points of view from which a field theory maybe regarded as "unified." (1) The field quantities should appear as unified, covariant entities which are irreducible under the invariant group of the theory. (For instance, the electric and magnetic fields experience a unification in this sense under the Lorentz group of special relativity), (2) The Lagrangian of the theory should not be expressible as the sum of several invariant parts, but should be a formally unified entity.[75]

In the latter case the total range of properties of physical objects would be correlated; in the former case different properties or groups of properties would give rise to separate contributions to the total geometry, or rather, to the geometries of space-time.

The central element in all this movement is the search for suitable invariants, the invariance being the basic requirement of the general principle of relativity. Moreover, that search for invariants is recognized, even if reluctantly, as a search for the explanation of the physical properties of things. The reluctance finds different expression on different levels. Thus, on the theoretical level, D. G. B.

Math. Phys. (6) 1965, 94–102, 'Invariant Approach to the Geometry of Spaces in General Relativity,' where a procedure is described for obtaining a complete, invariant classification of the local analytic geometries and matter fields in general relativity by a finite number of algebraic steps.

[73] Veblen, op. cit., 1927, 33.

[74] Cf. the references in footnote 72.

[75] W. Israel and R. Trollope, 'New Possibilities for a Unified Field Theory,' *Journal of Math. Physics* (2) 1961, 786.

Edelen bases his advocacy of a theory of variant fields implicitly on such a reluctance, on the view that

> a field has no *a priori* existential properties in itself, but rather is an abstraction arrived at by a classification of the types and nature of interactions between matter and something we term a field. Thus, in order to satisfy the principle of observational covariance, we need only to require that the equations describing matter and the interactions of fields with matter be covariantly formulated, but not that the field equations themselves have such properties.[76]

C. W. Misner and J. A. Wheeler contrast interestingly two opposing views on the nature of physics:

> (1) The space-time continuum serves only as *arena* for the struggle of fields and particles. These entities are foreign to geometry. They must be added to geometry to permit any physics. (2) There is nothing in the world except empty curved space. Matter, charge, electromagnetism and other fields are only manifestations of the bending of space. *Physics is Geometry.*[77]

While the first view comes close to the commonsense reluctance which will concern us presently, the second view—the one enlarged on in the paper cited—is expressed even more unhappily than that of Veblen cited above.[78] It is, however, close to the position we have exposed.

On the popular level the reluctance we spoke of is entrenched, for on that level neither the significance of the invariance of special relativity theory nor that of the general theory is acknowledged. Invariance is admitted on the levels of chemistry, biology, etc., for, the acceptance of invariance on these levels is a spontaneous acknowledgment of an empirical residue. But the acceptance of invariance in physics, since physics deals precisely with local motion, calls either for the inverse insight of special relativity which rejects inertial transformation as a significant factor in explanation, or for the difficult and more basic insight of general relativity which establishes the unrelatedness of the viewpoint of the observer to the properties of things.

This reluctance on the popular level, and on the level of commonsense philosophy, is related not only to a lack of understanding of science but also to the peculiarities of knowing particulars. This latter problem is one which is a fundamental source of difficulty to general relativity physicists at present, especially in relation to the problem of the integration of quantum theory and

---

[76] *The Structure of Field Space*, Berkeley, University of California Press, 1962, 36–37.

[77] 'Classical Physics as Geometry: Gravitation, Electromagnetism, Unquantized Charge, and Mass as properties of Curved Empty Space,' *Annals of Physics*, (2) 1957, 526.

[78] Footnote 73.

general relativity.[79] These difficulties are expressed extremely honestly and enlighteningly by E. P. Wigner. He remarks that while special relativity and quantum theory have to some extent been brought together,

> this is not so with the general theory of relativity. The basic premises of this theory is that coordinates are only auxiliary quantities which can be arbitrary values for every event. Hence, the measurement of position, that is, of the space coordinates, is certainly not a significant measurement if the postulates of the general theory are adopted: the coordinates can be given any value one wants. The same holds for momenta. Most of us have struggled with the problem of how, under these premises, the general theory of relativity can make meaningful statements or predictions at all. Evidently, the usual statements about the future positions of particles, as specified by their coordinates, are not meaningful statements in general relativity. This is a point which cannot be emphasized strongly enough and is the basis of a much deeper dilemma than the more technical question of the Lorentz invariance of the quantum field equations.[80]

The deeper dilemma is closely related to the question already discussed: the concern of classical science with primary relations, the concern of statistical science with general knowledge of secondary determinations through distributions. These secondary determinations are reached only through particularization, whereas the invariance of general relativity breaks down under particularization.[81]

Moreover, particularization does not involve an explanatory addition. On the refined level of science this relates, for example, to the dependence of quantum theory on classical science for its definitions. On the more general level it relates to the identification of systematic explanation and invariance with the properties of things, properties which of their nature are common. To pin down the individual one must move from the region of invariance and explanation to the region of ostensive definition and particular reference frames. Finally, returning to Kneale's problem, we may claim that specifying the ultimate alternatives is a contradictory notion or project if the ultimate alternatives are individuals and specifying means conceptually or systematically distinguishing the individual uniquely. Firstly, Strawson's work gives the lines of the unique distinction of individuals. Secondly, an individual may well be conceptually

---

[79] Cf., for example, John Boardman, 'Contribution to the Quantization Problem in General Relativity,' *Journal of Math. Physics* (6) 1965, 1701.

[80] 'Relativistic Invariance and Quantum Phenomena,' *Review of Modern Physics*, (29), 1957, 255.

[81] Cf. Lindsay and Margenau, *Foundations of Physics*, 368.

unique, but this uniqueness will be factually ascertained, and the uniqueness will be the uniqueness of the particular set of common properties possessed.

We may return now to the question of lacunae in Aristotle's view, and to the problem of a more adequate definition of chance. Adequacy here is not just a matter of harmonizing with ordinary usage as Kneale might claim:[82] one must rather

> delimit a subject matter which possesses a number of characteristics considered to be particularly important to it and which form a meaningful unity; the "correctness" of this concept can only be judged from its fruitfulness, both as a creative idea in life and as a useful instrument for advancing scientific investigation.[83]

The clue to that adequate definition lies in Aristotle's comment that 'to say that chance is a thing contrary to rule is correct.'[84] For Aristotle, the domain which lay 'outside rules' included all that could not be 'regulated' by some form of classical, causal, law. There was for him no scientific law concerning death by falling slates in Athens or through horse kicks in the Prussian army. Since Aristotle's time his view of chance has dominated, explicitly or implicitly, other discussions of chance. It was not replaced by a more adequate account after the emergence of a science of statistics, and so there arose the paradox of which Poincaré speaks. His quotation from Bertrand's *Calcul des Probabilités* sums up the matter: 'How dare we speak of the laws of chance? Is not chance the antithesis of all law?'[85] Law means rule: chance means being contrary to rule. To solve the paradox one must acknowledge that what we now know as laws of chance are laws of what Aristotle called chance when he discussed what in his time was 'outside rule.' If we are to employ the word 'chance' now with Aristotle's general meaning 'contrary to rule' we must employ it in regard to a more restricted domain than Aristotle did. In the process of specifying this restricted domain we will answer the question, Why are there laws of what Aristotle called 'chance'?, a question which is obviously closely related to Kneale's problem: 'Why should we find order on a large scale combined with disorder on a small scale?'[86]

Why, then, are there laws of probability relating to, say, death through horse kicks in the Prussian army? To remark that 'the simplicity of the result is born of the very complexity of the error'[87] or of the causes, as Poincaré does, is to fail to get to the heart of the matter. Nor is it enough to say that the group's behaviour

---

[82] Probability and Induction, 169–170.

[83] Werner Sombart's comment on his attempt to define his subject, *Proletarian Socialism*, as quoted by von Mises, *Probability, Statistics and Truth*, 3–4.

[84] *Physics* ii, 5, 197a, 18.

[85] *Science and Method*: English translation from The Foundations of Science, 395.

[86] *Probability and Induction*, 164.

[87] *Science and Method*: trans, as in footnote 64, 406.

is predictable '*precisely because the odd things that one individual does will tend to cancel out the odd things some other individual does.*'[88]

Consider the well-known example relating to heights in a human population. Let us do what Aristotle never thought of doing with the men of Athens: let us take an adequate sample—and we will see below what problems this phrase cloaks—from, say, a list of heights. When we graph the results in the usual way we have a series of points, which lie, roughly, on a normal law curve. For our argument it is in fact strictly necessary only that we have a set of points on the number-height plane, however disorderly, but one could also discuss why the set of points turn out as they do, lying so as to suggest a continuous curve. Basically, it is because there is *no reason* for discontinuity, or for peculiar irregularities: 6 feet tall is not significantly different from $6 + \frac{1}{100}$ feet tall. These comments are of course *a posteriori*. We are not deducing that population heights obey a certain type of law: as von Mises repeatedly remarks,[89] probability hypotheses are no different from other scientific hypotheses as regards their origin and structure. As we shall see, his remark holds true even as to their relation to the fundamental principle of induction.

Next we take another 'adequate sample' and repeat the process, getting another set of points on the graph. Our question now is, what is the relation, if any, between the two sets of points? Will they be significantly different in their distribution: will, for example, their 'centres of gravity' be significantly removed from one another? Our experience of sampling of course provides the answers but we seek now an understanding of that experience. What, for instance, is meant by 'significantly different'? We recall Mill's discussion of a not unrelated problem:

> If after such a number of trials that no further increase of their number has any material effect upon the average, we find a preponderance in favour of a particular throw (with a die), we may conclude with assurance that there is some constant cause acting in favour of that throw, or, in other words, that the dice was not fair; and the exact amount of the unfairness.[90]

Since we assumed that we had two adequate samples, that is, among other things, 'such a number ... that no further increase ... has any material effect,' a significant difference will be one signifying a 'more than random' difference in the two adequate samples. But who is to judge the number which is sufficient, the difference which is significant?

---

[88] R. G. Lipsey, *An Introduction to Positive Economics*, London, Weidenfeld & Nicolson, 1966, 13.

[89] *Probability, Statistics and Truth*, esp. 30 ff.

[90] *A System of Logic*, 350.

The most intimate evidence of randomness is the judgment of the experimenter. The man familiar with the technological aspects of the articles under consideration is in a position to be a good judge—perhaps the only competent judge—of whether the selection was ... unbiased with respect to the quality characteristic under consideration.[91]

This does indeed seem to leave us very far from an answer to our questions and from the meaning of 'significant difference' or 'adequate sample,' and without this meaning it is not easy to understand why differences between samples should be of interest at all. We may perhaps say that a sample is adequate when it is 'similar' to the population, when it is not 'significantly different' from the population, and if this is true for each sample obviously the two samples would not be expected to differ significantly. But are we not arguing here in circles?

At the foundations of any science there must be some type of 'circular,' explanatory or implicit, definitions which can be cast into axiomatic form. Operative in the sampling process generally, and in the preceding paragraphs in particular, have been such basic axioms, in an implicit, 'uncrystallized,' form. To recall a parallel already made, the preceding paragraphs resemble the solution of a problem on Euclidean geometry by diagram, in which Pasch's axiom of order or indeed other axioms of the geometry are implicitly assumed. The special axioms operative in the present context would undoubtedly be some form of the axiom of randomness and of the axiom of limiting frequencies. The procedure and results of sampling clearly involve both these axioms. But there is another basic axiom involved here, an axiom which, as fundamental axiom of empirical science, statistical science shares with ordinary causal explanation: it is the axiom that 'similars are similarly understood.' Thus, as we implied above, to say that a sample is not significantly different from the population is to say that it is similar to the population. The meaning of 'similar' in the case of statistical science is specialized by its coupling with the two von Mises, axioms. If two samples and the population are all 'similar'—where this 'similar' in fact has reference more to description—then one expects that they be similarly understood. The general axiom underlies not only Mill's reasoning in the quotation above but more generally his four methods of inquiry. It is the basic axiom of induction. For the empirical scientist, the real problem of induction is the problem of criteria of relevant similarity on the descriptive level. He has no problem of applying his theory to a similar instance provided he can judge the instance to be relevantly similar, and in fact no amount of formulated criteria can replace his judgment based on experience and experiment.

---

[91] Leslie Simon, *Engineers' Manual of Statistical Methods*, New York, John Wiley & Sons, 1941, 163.

When we restore to our concept of sampling the normally accompanying language of "how to go about it," it becomes evident that the reliability (rather than probability) of sampling inferences springs from the watchful labours of the sampler, not from the application of a calculus which takes no account of those labours. If someone were to object that no pains short of complete enumeration in the smallest degree force a sample *logically* to match a population, I should want to answer that the logical necessity he speaks of is in another world from sampling.[92]

The other world would seem, indeed, to be a favourite haunt of many, such as Goodman, and we cannot but agree with R. J. Butler's comment on Goodman's Riddle: 'there are enough problems surrounding the notion of class without our imagining that perverse predicates like "grue" pose more.'[93] If it is the sayings of scientists that are the interest of the philosopher of science, then a much better candidate than green and grue for discussion in this type of induction problem would be what the scientists wrote and said of the process of passing from a definition of water to the definitions of ordinary and heavy water.

With these background problems clarified we come nearer to the solution of the question, why order is found in disorder. The disorder in which statistical science finds order is the absence of systematic relations between the secondary, concrete determinations of classical laws. This man's height is determinate; that man's height is determinate; but there is no systematic relation between their determinate heights. There is obviously *some* relation of proportion between their heights, being quantities, but that this man is one inch taller than that man is just a matter of fact which is not to be systematized. Similarly, the list of heights in a population, or in a sample, is non-systematic. So, our question of order from disorder may be more precisely formulated as, why should law or system be associated with absence of system, with what is non-systematic?

Let us return to our principal example and ask, Why should each adequate sample be nonsignificantly different in distribution from every other adequate sample? In other words, why should regularity enter into the comparison of coincidental aggregates of this type? Because if the differences were not just random or nonsignificant, we would suspect the adequacy of the samples or we would suspect that there was some systematic interference. By an odd twist, the twist indeed of an inverse insight, the regularity that enters in can be appreciated as a measure of just how non-systematic the samples are. As Mill remarked regarding the biased die, 'if ... we find a preponderance in favour of a particular

[92] H. A. Nielsen, 'Sampling and the Problem of Induction,' *Mind* (LXVIII) 1959, 481.

[93] 'Messrs. Goodman, Green and Grue,' in *Analytic Philosophy: Second Series*, ed. R. J. Butler, Oxford, Blackwell, 1965, 193.

throw, we may conclude with assurance that there is some constant cause acting in favour of that throw.'[94] The sample distribution with this die differs from the sample distribution of other dice and so we 'conclude with assurance' that there is a systematic factor entering in. Our 'assurance' springs from the fact that we do not expect divergence from the non-systematic to be systematic. In the population example, one sample was, by assumption, as non-systematic as the next. Each had *some* distribution associated with it. If the samples are genuinely similar in their non-systematic nature, then there cannot be a systematic divergence between their distributions. So, the distribution of *any* adequate sample sets a norm from which distributions in other adequate samples cannot diverge systematically. And if one takes a limit in some way or other—the difficulties of this will concern us in chapter eight—one may arrive at a continuum of cluster points, ideal frequencies, such as those given in our example by the normal law, which provide a norm from which adequate sample distributions can only diverge randomly, or non-systematically. A systematic shift, for example to a greater mean height, would connote the interference of a systematic factor, like an improvement in general diet. In general terms then, a probability distribution gives an ideal norm from which concrete non-systematic measurements can diverge only randomly.

This brings us also to the solution of Kneale's first question regarding the meaning of chance. Aristotle recognized the non-systematic which resulted from the coincidence of causal sequences. It lay outside rules and so qualified for the name 'chance.' Modern science has brought that non-systematic to some extent under rule, but not entirely so. For, while concrete results cannot diverge systematically from correct probabilities, still they can diverge non-systematically or randomly. There are no rules governing these random divergences of measured frequencies from the ideal frequency distribution given by probability theory. There seems, therefore, no better candidate for the title of chance in contemporary science then the non-systematic divergences of actual frequencies from the ideal frequencies called probabilities.

Against this background we may now comment more adequately on the notion of chance as it occurs in evolution theory and in certain relevant discussions of modern physics.

David Bohm, in discussing the indeterminist philosophy, remarks that we might use the name 'absolute chance' in regard to the view which would hold that there are irregularities which cannot in principle be accounted for by microphysical parameters, and that these irregularities 'could not have the character of ordinary chance fluctuations, which represent the effects of contingencies that cannot be taken into account in the context under discussion. Rather, they would represent a kind of fundamental and irreducible arbitrariness

---

[94] *A System of Logic*, 350.

or lawlessness in the detailed behaviour of the world.'[95] Here is not the place for a detailed discussion of the indeterminist position: we are interested more at present in what meaning the notion of 'absolute chance' could have when considered in the light of our own treatment. On our view, chance is a type of residue associated with law. In so far as world process has not the systematic perfection of a theoretical planetary system, empirical science will have to admit such a residue. The absolute chance of the indeterminists is not, however, residual in this way. It is fundamental: as Bohm puts it, it replaces the idealized frictionless machine of Laplace by 'an idealized roulette wheel that would give an irregular distribution of results depending on nothing else at all.'[96] The best that can be exacted on this foundation of ultimate lawlessness is a probability-type theory, so that *'laws of probability are regarded as having a more fundamental character than is possessed by determinate laws,'*[97] and so it becomes a commonplace to say that 'One of the most important insights contributed by the theory of quantum mechanics is this: The fundamental laws of nature are laws of probability, not laws of certainty.'[98] From there it is not a long step to the conclusion that classical laws are something of a macroscopic illusion, an epiphenomenon.

With regard to this last point we may say immediately that its full consideration will occupy us later.[99] Suffice it to note here first of all that on any adequate view of verification the laws of behaviour of the elephant are at least as well verified as the probability laws of electrons. Further we note that if classical laws are not verified by identity with the concrete, neither are probability laws: both types of law are ideal; from both, the concrete processes and events diverge though in different ways. Thirdly, our earlier[100] considerations of the dependence of statistical laws on classical laws for definitions of states etc. expose a basic weakness in the view that 'the fundamental laws of nature are laws of probability, not of certainty.' One has the results of an idealized roulette game, a distribution of states, but how are these states classified, distinguished or defined? What is involved at any level of physics is not simple causal considerations nor pure probability considerations but a complementation of one by the other in reaching some understanding of the aggregate of processes at that level. Nor is this a matter of *a priori* reasoning: the textbooks and the experimenters are on our side. Since both methods are involved in the investigation of any level, the suggestion of alternating statistical and causal theories descending indefinitely on to a terminal theory of either type is remote from actual scientific practice. Again, empirical methodology has nothing to say about the number of possible revisions or levels:

---

[95] David Bohm, *Causality and Chance in Modern Physics*, 62–3.

[96] Ibid., 63.

[97] Ibid., 64; italics his.

[98] K. W. Ford, *The World of Elementary Particles*, New York, Blaisdell, 1963, 51.

[99] Cf. chapter 9.

[100] Cf. above, p. 64.

it can anticipate only the method of approaching such revisions. Finally we note that besides the 'unlawfulness' of chance which we made precise above, we also admitted a general basic residue, the empirical residue,[101] which in fact resembles the 'kind of fundamental and irreducible arbitrariness' that Bohm associates with 'absolute chance,' or the notion of 'pure chance' or 'absolute disorder' which W. R. Thompson describes:

> A universe of pure chance is, in the strict sense of the word, unthinkable, by which we mean, not something distasteful or dissatisfying, but something on which the mind cannot take hold at all. A world of pure chance is simply chaos, or *absolute* disorder, and the concept of absolute disorder has no positive intelligible content.[102]

Our empirical residue was defined precisely as something which had of itself no positive intelligible content, but which, as it were, showed up in its relation to other factors of scientific importance. Thus with the empirical residue can be associated the occurrence of identical situations and processes; with what is presupposed in any geometrical ordering; with the basic assumption of scientific generalization and collaboration; with the possibility of conjunctions of causal chains discussed by Bunge; with the occurrence of non-systematic processes which fall under statistical method but leave a residue of 'chance.' While the empirical residue is related in various ways to different aspects of scientific method, it itself is implicitly taken as 'irreducible' to lawfulness. Unlike 'absolute chance,' however, it has not been deduced from a philosophic position: it has been admitted as an unavoidable factor in empirical investigation.

There is another relevant usage of 'chance' which can be exemplified from the book of K. W. Ford already quoted.[103] Thus, it is stated that 'we seem to see the hand of chance working at every turn—chance that any particles are stable, chance that the neutron can live forever within the nucleus, chance that we are free of the threat of annihilation by antiparticles.'[104] Again, 'what stabilizes the neutron is a rather peculiar "chance," that the pion exchange force between a neutron proton happens to be somewhat stronger than the same force acting between two protons,'[105] 'the key fact is that it is only "by chance" that the electron lives forever.'[106] The quotations reveal an interesting aspect of empirical method: it is, that the scientist is concerned with what happens to be, not with what necessarily is so. The pion exchange force and the electron lifetime might,

---

[101] Cf. above, pp. 22, 37, 90.

[102] *Science and Common Sense: An Aristotelian Excursion* (London, Longmans, Green, 1937), 218, quoted in Fothergill, *Evolution and Christians*, London, Longmans, 1961, 309.

[103] Cf. footnote 98 above.

[104] Op. cit., 237.

[105] Ibid., 174.

[106] Ibid., 161.

so to speak, have been otherwise, but as a matter of fact they happen to be so. The empirical scientist attends to what *de facto* is given, and the hypotheses which he formulates about that given express possibilities, never necessities. The use of 'chance' in contexts such as the above, while somewhat confusing, emphasizes the point. The long dispute over the Euclidean parallel axiom is another aspect of the same point: the emergence of other geometries exposed its non-necessity and it is now seen to be a *de facto* matter that Euclidean geometry is physically verified as a good approximation. Later we will come to see this from another point of view when we come to discuss the deficiencies of definitions of empirical probability.[107] The scientist, then, is restricted to the explanation of processes and occurrences which he accepts as facts. In so far as he is tempted to make that restriction a more general restriction he is no longer engaged in science but in poor philosophy.

When we turn to evolution theory we find that the notion of chance is involved mainly in two contexts. First of all there is the context specified by Fothergill in the following passage:

> Natural selection could be said to be the cause of evolution. According to this theory evolution seems to occur by means of a haphazard process. Hence the phrase "evolution occurs by chance." Now "chance" is technically a mathematical word, and as such it does not mean that the operations concerned in it are working against or without known laws.[108]

Fothergill is correct in pointing out here that the 'chance' of this context is the chance of 'Laws of Chance.' Against the background of our discussion of the paradox of having laws of chance, we can see that it is more correct to speak of evolution as 'being governed by probabilities' rather than by chance, though even this is a doubtful expression and in a later section[109] we will show more exactly how natural selection is given scientific expression through a statistics of survival.

The second context in which 'chance' occurs in discussions of evolution theory is that involving the expression 'chance variation.' While in this field of investigation there are finer points of discussion, such as the question of genetic assimilation, it is evident that the basic variations relevant to the evolutionary process are on the level of the 'genes.' The causes of mutations, translocations etc. on this level, far from being nonexistent, are a continued object of investigation.[110] These mutations are relevant to the scientific consideration of evolution: but what can be meant in this context by 'chance mutations'? What is in fact relevant is the viability of mutated genes, where 'by viability is meant the

---

[107] Cf. chapter 8.
[108] *Evolution and Christians*, 43.
[109] In chapter 11.
[110] Cf., e.g., the discussion and references in Fothergill, op. cit., 178–194.

successful negotiation by the mutant of the whole internal and external environment in relation to like and unlike mutants and species also present.'[111] Experimenters have in certain cases provided figures both for frequency of occurrence of mutations under particular conditions, for example, radiation, and for frequency of survival of the resulting mutated plant or organism. The latter frequency, the frequency of survival, is clearly more related to natural selection as we discussed it above. The former frequency, the frequency of emergence of mutations, is the aspect of 'chance variation' which is of importance in a scientific study of evolution. What is significant for that theory is the probability of emergence of various combinations of mutations. The random divergences from such probabilities are not scientifically relevant. Experimental work in this region aims at providing the probabilities and allowing for the divergences. These random divergences would therefore come under our definition of chance. The phrase 'chance variation' and its treatment in discussion involves in fact the paradox which has been our concern right through this chapter.

---

[111] Fothergill, op. cit., 181.

As our title involves us in three topics, randomness, statistics and emergence, so the present discussion treats of the question of verification in relation to these three topics. In chapter three our attention centred on the question, Is there real randomness J, and our affirmative answer implied a view on verification and objectivity which called for fuller expression, which we now undertake. In chapter nine the problem of real emergence arises, and chapter eleven concludes with a discussion of the problem of the verification of an evolutionary hypothesis. Here we look forward to these chapters but we will be dealing more directly with the verification of statistical hypotheses, and that in the context of the more general problem of verification.

As we shall see, the specific difficulties of probability theory, and of the two types of probability, help towards a clarification of the nature of verification in general. Thus, as Dr. Marjorie Grene points out, R. B. Braithwaite takes as the pattern for scientific laws Hume's constant conjunction model and shows with great care and precision how the deductive theories of science can be cut to fit it.[1] But when he comes to probability hypotheses he is confronted with the peculiar problem they present to the constant-conjunction empiricist. This peculiarity of probability statements is due to 'the fact that they occur in deductive systems which are fitted to experience in a looser way than are deductive systems in which probability statements do not occur.'[2] The consequent investigation leads Braithwaite to strange conclusions:

> The question of the reasons for the acceptance of any hypothesis, whether statistical or universal, will be found, in the last resort, to involve teleological considerations. ... The peculiarity of statistical reasoning is that it presupposes also at an early stage of the argument judgments as to what sort of future we want. In considering the rationale of such thinking we cannot avoid ethics breaking into inductive logic.[3]

This intrusion of ethics into scientific knowing is made much of by Grene, who follows M. Polanyi in considering an evaluative element to be an essential factor in knowledge.[4] No one could regard this as a happy solution to the problems

---

[1] M. Grene, *The Knower and the Known*, London, Faber, 1966, 114118; R. B. Braithwaite, *Scientific Explanation*, 10.

[2] R. B. Braithwaite, op. cit., 131.

[3] Ibid., 173–4.

[4] M. Grene, op. cit., 157–182; M. Polanyi, *Personal Knowledge*, London, Routledge, 1958, chs. 8, 10.

related to the constant conjunction view and to verification. We may recall Hume's remark on his position: 'For my part, I must plead the privilege of a sceptic, and confess that this difficulty is too hard for my understanding. I pretend not, however, to pronounce it absolutely insuperable. Others, perhaps, or myself, upon mature reflection, may discover some hypothesis that will reconcile these contradictions.'[5] We will return later to the attitude implicit in this quotation. Our hope is of course that the discussion of the problem of the verification of statistical hypotheses may open the way to the discovery of some hypothesis to solve Hume's difficulty.

Our interest here, no less than elsewhere, is in the process of discovery: for verification, or falsification, is no less a discovery than is the arrival at a possible theory. It is a different type of discovery in that it is an answer to a different type of question. In the last chapter we remarked on the lack of emphasis in Aristotle on the problem of verification. Nonetheless, his list of questions mentioned there clearly includes two types of question. There is the what-or why-question, with which we mainly dealt in the last section, and there is the is-question, 'is it so?,' 'is it the case?,' which is our present concern. It is the is-question which drives the relativity physicist halfway round the earth to observe an eclipse. It presupposes some answer to the what-question—the eclipse observer has at his fingertips both theory and expected sensible consequences—and its own answer adds nothing more to the what-answer than a Yes or a No. Theoretically the procedure of verification adds no new meaning, but in practice this is by no means normal. Thus, the jury on retiring to proceed to answer the usual is-question is in possession of all the facts and clues but frequently it is through the effort to find an is-answer that the evidence comes to be properly understood. The example recalls Toulmin's aim in his book, *The Uses of Argument*, that 'the nature of the rational process will be discussed with the "jurisprudential analogy" in mind.'[6] We have obviously a common interest, but here the field of argument or verification is the special field of statistical hypotheses. The problems referred to in Toulmin's introduction are manifest in this field as in the broader field of Toulmin's discussion. The focus of the problem in our particular field is neatly described by H. A. Nielsen as

> the belief that we can capture the "logic" of sampling just as we capture the logic of Barbara: in abstraction from all matter affecting the *conduct* of the inference. Once this is supposed, it is no step at all to imagine that the strength of sampling lies somehow in the formula, as is true of Barbara, or else in cumulative applications of the formula. To be carried along in this way, however, is to fail to see that, while the paradigm of Barbara *prescribes* the conduct of the inference, *viz*, by

---

[5] Hume, *Treatise of Human Nature*, Oxford, 1896, 636.
[6] S. Toulmin, *The Uses of Argument*, Cambridge University Press, 1958, 8.

substitution, the paradigm of sampling is altogether useless for making a good sampler out of a duffer.[7]

The logic of discovery in the verification of statistical hypotheses is by no means a neglected area—witness the dense central section of Braithwaite's *Scientific Explanation*—but with rare exceptions such treatments tend to become remote from scientific practice and the foundation problems of actual science. Thus Braithwaite, in the early sections of the book mentioned, bases his discussion on a theory manufactured for the purpose, and in the section of interest to us, to quote Harold Jeffries, 'the method used for assessing probabilities would lead to appalling complications if it was applied to a continuous distribution, or, still worse, to a stochastic process.'[8] Again, Spencer Brown, in his *Probability and Scientific Inference* is so carried away by his thesis of the meaninglessness of the concept of statistical probability, carried away too from the ordinary practice of statistical science, that he can conclude to paradoxes of probability[9] which would cause little anxiety to the worker in the field.

For us, what is important is the ordinary practice of the statistical scientist in the process of verification. Moreover, it is that practice as it occurs in ourselves that is the basis of discussion: to discuss the practice without that experience would be as futile as doing biological research without specimens. That ordinary practice in its outline is familiar. One arrives at, or begins with, a particular statistical hypothesis. One gets a set of observation results. Do they fit the hypothesis? One knows—for one has what the ordinary scientist would consider to be an understanding of statistical hypothesis—that the hypothesis is not about any one set of observations, and that in fact the given set of observation results does fit the hypothesis if it has a probability of occurrence, no matter how small.[10] Still, the occurrence of a set with small probability—say Fisher's—would leave one uneasy; the occurrence of a second like set would leave one more than suspicious; etc., etc. We may say that it is the 'etc., etc.' that is important here, the manner in which the investigation is carried forward to a somewhat conclusive answer, Yes or No, to the question, 'Is this hypothesis verified?' Our efforts, then, are directed towards thematizing the scientific experience of verification, and drawing out its implied suppositions.

This attitude towards the problem of verification and objectivity, and particularly the stress on Questions, is not continuous with much of the

---

[7] 'Sampling and the problem of induction,' *Mind* (LXVIII), 1959, 479.

[8] In a review of Braithwaite's book, *British Journal for the Philosophy of Science* (4), 1953–4, 348.

[9] Esp. in chapter 9, op. cit.

[10] This is a problem with which we will deal in chapter eight.

contemporary discussion of these problems.[11] There is a large body of literature on Verification, and a wide range of views on the objectivity of theory, theoretical entities, etc. To review this corpus here would be space-consuming and superfluous. We may remark, however, at this midpoint of the presentation of our total view, that the position taken here serves to bring together, in the context of an understanding of actual scientific practice, what seem the best elements of the range of views. Thus, for example, one is enabled to see that there is room in reality for both atoms and Individuals, for both molecules and men. Both epistemological reductionism and ontological reductionism are excluded,[12] and the emergentist view justified. But the exclusion and the justification are not through detached metaphysical arguments: rather the argument is an explicitation of the implicit exclusion of reductionism by both scientist and 'man in the street,' and the justification is through an exposition of methodological suppositions. We will return to these issues later.

Perhaps we might make a start to our discussion from the commonly accepted fact, that a theory is not acceptable without some possible empirical checking. To say this is not to commit oneself to any particular view on the nature or objectivity of theory. One can admit it provided one has some view to that extent compatible with the empirical tradition. We may move from this starting point into our own line of investigation by throwing the stress on the checking, and on the questioning attitude of the scientist involved in the checking.

Verification does involve a return to sense experience, but the return is intelligent and within a context of theory. The verification of Einstein's general theory may lie in the observation of eclipse phenomena, but the relevance of the observation is appreciated only through the theory and the result is significant only within the theory. Moreover, it is not the observation but the intelligent observer that bestows some degree of probability on the theory, and it is for this reason that the maxim of La Rochefoucauld, 'Everybody complains of his memory but nobody of his judgment,'[13] rings so true. Verification is a question of intelligent judgment on the part of the verifier. The question which dominates the procedure of verification is the question, Is it so? Is the sodium D-line double? Is the distribution Poisson? This type of question involves a quite different mental stance from the what-question. It is this difference in mental stance which would seem to be the basis for Toulmin's general discontent in *The Uses of Argument*. It

---

[11] There is a growing interest in the question in various schools: cf. M. Novak, *Continuum* 2, 1964, 'The Performance of Asking Questions,' 389–401; E. Coreth, *Metaphysik*, Innsbruck, 1961; M. L. and A. N. Prior, 'Erotetic Logic,' *Philosophical Review*, 1955.

[12] Cf. E. M. Mackinnon, 'The New Materialism,' *Heythrop Journal* (8), 1967, esp. 11–14.

[13] Maxim 89 in F. Duc de La Rouchefoucauld, *Maxims*, trans. L. W. Tanock, London, Penguin, 1959, 46.

is this difference of mental stance which Newman stressed in his Grammar of Assent. It is the mental stance which Tarski finds so hard to appreciate:

> I do not have the slightest intention to contribute in any way to those endless, often violent discussions on the subject "What is the right conception of truth?" I must confess I do not understand what is at stake in such disputes; for the problem itself is so vague that no definite solution is possible.[14]

In Tarski's discussion of truth there is no question of criteria raised: what is at stake is a type of definition of truth, an answer to a what-question. The criteria of truth and their understanding involve the mental stance of the is-question which is in general understressed in so far as one's concern is entirely with a strict logic of verification. Such a concern can lead to a consistency theory of truth or to an elaboration of a logic of testing[15] which bears little resemblance to scientific practice. This is not to say that the apparatus of logic is irrelevant to verification. But the apparatus of deductive logic pertains rather to the what-question and to the what-answer which is a prerequisite of verification, of judgment. It thus contributes—as we shall see more clearly later—to a structured objectivity in which observation, coherence and critical reflection play each their essential part. Its contribution is to the synthesis which is prior to judgment or verification, to the formulation which is, so to speak, the borrowed content of the is-question. But there has been a general neglect of what lies beyond the synthesis, a neglect which has its roots in Aristotle[16] and which is evident in Kant: 'I find that a judgment is nothing but the manner in which given modes of knowledge are brought to the objective unity of apperception. This is what is intended by the copula "is."'[17] Through that neglect there is a stress on what pertains to synthesis, and the role of the copula 'is' in that synthesis. That synthesis is propositional, and provides an object of consideration which may be either affirmed or denied, but the important point is that in that affirmation or denial the copula 'is' occurs in a new sense, as answer to the is-question, Is it in fact so? Is it the case?

One might take as paradigm for critical reflection and for the transition from synthesis to positing of synthesis the jury room preoccupation with the question, Guilty or not guilty?, and the emergence of the verdict. As the jury, so the experimentalist must seek to be alert, free from bias, sufficiently well-informed

---

[14] 'The Semantic Conception of Truth,' *Phil. and Phenom. Research* (IV), 1944, reprinted in *Readings in Philosophical Analysis*, 64. Cf. also the dispute between Gonseth and Tarski at Brussels, *Internat. Cong. of Philos*, 1953.

[15] For a survey of such attempts cf. A. Pap, *An Introduction to the Philosophy of Science*, ch. 13.

[16] Cf. B. Lonergan, *Verbum: Word and Idea in Aquinas*, 47–94; CWL 2, 60–104.

[17] *Critique of Pure Reason*, B.142. One might trace modern parallels, e.g., K. Popper, *Conjectures and Refutations*, London, Routledge, 1965. 225ff.

not only about the focal question but also about fringe factors which may interfere. Certainly he must be anything but 'a duffer' in the field. He may well avail himself of prescribed procedures, such as that of maximum likelihood, but his actual procedures will bear little resemblance to the procedures advocated by Carnap or Wald or Braithwaite.[18] Ultimately he will make a statement—normally in a scientific journal—on his theory, his experiments, and the confirmation of the theory by the experiments. His statement, if adequate, will convey a sufficient indication of his certainty. As has often been noted, he is not absolutely certain, but he claims that the theory is adequately verified, it is probable or highly probable. The probability in question, however, has nothing to do with the probability which we have discussed so far. In this matter Carnap is correct[19] and we might follow his subscript method for distinguishing the two probabilities. We find it more convenient, however, to use the term v-probability for the probability of confirmation and f-probability for the probability special to statistical science.

There are various reasons for the confusion between these two types of probability. The first source of confusion is the neglect, already considered, of the different mental stances of the two types of question. A further source of confusion is the fact that statistical probability centres its attention on occurrences and events, and these are known by answering is-questions. Another source of confusion is the fact that certainty is not a black and white affair: it is a spectrum. We will discuss in turn these sources of confusion.

The first and basic step is to distinguish between probability as an answer to a what-question and probability as a quality of an is-answer: the former is f-probability, the latter is v-probability. To distinguish them, as we shall see, is not to separate them: they are in many ways interlocked. But without the distinction, which is based on the two different mental stances, the character of their interlocking would be permanently obscured. We have already seen an instance of this interlocking through the discussion of Reasonable Betting and what follows will throw further light on the conclusions reached in that chapter. The stance of the man making a bet is one of judgment. In so far as he hopes to move beyond a guess he must weigh the evidence in favour of the prediction 'A will win.' The context of the prediction may include statistical information of different kinds, answers to various what-questions, and, as we have already seen in chapter five, that information can be peculiarly availed of in reaching an answer to the question, Will A win? Still, the bet, in so far as it is solely based on statistical information, resembles a guess rather than a probable judgment. Both a guess and a probable judgment are based on incomplete knowledge: intelligent

---

[18] Cf. A. Pap, op. cit., for references.

[19] 'The Two Concepts of Probability,' *Readings in the Philosophy of Science*, edited by Feigl and Brodbeck, New York, Appleton-Century-Crofts, 1953, 438–455.

reflection in either case shows that evidence is insufficient for certainty. In the case of the probable judgment that insufficiency is partial: there is some approximation towards sufficiency which can be grasped as such, leading to the modest commitment of a probable judgment, a judgment which is probably true, which converges in a nonstatistical sense on true judgment. But the pure guess goes beyond the evidence: there is no reason to expect heads, yet one calls heads, and the latter illustration gives the clue to the sense in which one might say that a guess is probably true. Whereas a probable judgment is v-probably true, a guess has only f-probability. Guesses are 'probably true' only in the sense of diverging non-systematically from true judgments about the events in questions.

Before venturing further in discussing probable judgments it seems best to try to eliminate the second source of confusion of the two types of probability mentioned above. This was, the fact that f-probability has to do with events. In dealing with this difficulty we would hope also to clarify an issue raised before, namely, Of what is f-probability a property? Already we have seen that the majority of philosophers seem to favour the view that somehow f-probability is a property of pennies, dice etc. In the present section we will justify our rejection of that view.

Instead of asking, Of what is f-probability a property?, we would rather ask, of what is f-probability an understanding? For, to ask directly the former question is to set the investigation off on the wrong foot: it presupposes that what can be understood must necessarily be a property of something. The new form of the question makes no commitment on this. It does not exclude the possibility of f-probability being a property of something, but neither does it suggest it. It will help, perhaps, to begin by stating simply that f-probability is an understanding of events, of aggregates of events, or more accurately it is the formulation of that understanding. To reveal what is meant by this claim one must answer the question, What is an event? We will answer that question in a manner calculated to throw some light on the discussion of this entire chapter by saying that events stand to is-questions as properties stand to what-questions. However simple this answer, it has, as we shall see, far-reaching consequences. In it are the seeds of the solution to the problem of objectivity underlying our present discussion.

Already in chapter six we related properties to what-questions. It is quite evident that one cannot intelligently answer the normal what-question with a Yes or No: one ultimately answers with some type of correlation, or function, or law, or description. Yet the what-question is evidently related to the is-question. If one asks, What is a unicorn?, one can also ask, Are there unicorns? but only when one has some answer to the first question. If one asks, What is Piaget's or H. S. Sullivan's theory of the development of the child?, one will also be interested in knowing, Is one or the other verified? One cannot ask an is-question in the absence of any what-answer. Wherever there is question of verification or existence or occurrence there must also be some specification of what is verified or to be verified. On the other hand, one cannot rest content with a what-answer: one moves on to raise the is-question of verification. As the two types of question complement one another, so do properties and events, and it is through that complementarity that the meaning of 'event' is understood.

Now the source of confusion that we are speaking of comes from the fact that we can formulate what-questions about is-answers. That formulation, in everyday affairs, can take the form, How many?, and the answer is got by counting the events or occurrences, cattle in a field or eggs in a basket. In statistics the question has the generic form, How often?, and the answer here involves counting too, but a counting which yields not an exhaustive catalogue of numbers of kinds of events, but, through a statistical hypothesis, an ideal frequency of events. We have already discussed how this ideal frequency is reached from aggregates of events: here the aspect of importance is the relation of the formulation to events. It is an aspect indeed which has already been touched on in chapter five when we pointed out that a statistical hypothesis, like a classical hypothesis, could provide the middle term of an Aristotelian scientific syllogism. Can we say, then, that the formulation, the ideal frequency, is a property of aggregates of events? We might concede the usage, yet it would seem an unnecessary confusion, not because the frequency is ideal—with this problem we will deal immediately—but because the word 'property' is best restricted to specifying the object of investigation of classical science. But certainly the ideal frequency is not a what-answer, a property, of what is used to generate the aggregate of events.

Moreover the restriction in usage of the word 'property' is warranted by the difference in the verification of hypotheses in the two types of investigation. What statistical science gives is not the actual frequency of events but an ideal frequency of events. In what sense can we say that such ideal frequencies are verified? Certainly the verification in statistics differs from that in classical science—we recall Braithwaite's 'looseness of fit.' In classical investigation a law is tested by the substitution of numerical values determined by measurement for the variables that are functionally related by the law. But in statistics there is no possibility of the deduction of a determinate set of values with which to check particular experimental results. Whereas in classical investigations limits can be assigned to the divergence between measured and theoretical values, in statistics one must acknowledge the possibility of exceptional results. Still, experimentalists are well aware of these difficulties and they deal with them, not by having recourse to something like Braithwaite's hyperclasses,[20] but by a combination of well-tried techniques and informed intelligence. The basic difference in attitude between the classical investigator and the statistical investigator is expressed by saying that the former expects experimental results to converge on the proposed law,

---

[20] *Scientific Explanation*, chs. 6, 7.

whereas the latter expects that experimental results will diverge from the proposed law, but not systematically. The significance of this in practice for the dovetailing of the two methods has already been discussed in chapter five. But there is the basic similarity in the attitude of the two types of investigator to their results: both can lay claim to the verification of their theories with a certain degree of v-probability.

The last source of confusion of the two types of probability that we indicated was the fact that both probabilities admit, so to speak, of degrees. The degrees involved in v-probability are degrees of certainty, and these degrees are not fixed in neat categories such as some philosophers would have it when they speak of moral, physical and metaphysical certainty. They lie in a spectrum, and it is this spectrum that would incline e.g., Reichenbach[21] to consider that values of probability ought to take the place of the two truth-values of ordinary logic giving a multi-valued logic, or lead Carnap to consider a spectrum of v-probabilities as estimates on given evidence for a hypothesis.[22] The tendency in such approaches to overlook the mental stance involved in verification has already been indicated and criticized: the 'logical side of confirmation'[23] leaves out the essential. Still, the grounds for giving the probability of judgment or verification a numerical value are evident, and the question deserves closer consideration.

Braithwaite and Hempel and many others would consider universal judgments to be limiting forms of statistical judgments with p = 1 or p = 0. This is a more extreme view which adds earlier confusions to the question of estimation, but it helps us plunge straight into the difficulty by asking, What is the difference between the truth value 1 and unit probability? Firstly let us consider what the truth value 1 means. The truth value 1 normally occurs in the context of deductive logic: the truth value 1 is there given to e.g., tautologies, analytic propositions evidently acceptable from the meaning of the constituent terms or propositions and the rules of combination of such terms or propositions. On a larger scale the truth value 1 may be given to a syllogism or to an entire deduction or deductive system—wherever, indeed, consistency has been made evident. In such cases the focal question is the is-question, Is it consistent?, and the answer—in so far as the principle of excluded middle is applicable—is, Yes or No, with truth value 1 or 0. More refined treatment of this problem would carry us too far afield here. Again, a more subtle discussion would include a treatment of the meaning and factual reference of multivalued logics. On the question of factual reference one may note the parallel between the series of more refined logics and the developing series of geometries or the series of probability theories required

---

[21] *Experience and Prediction*, University of Chicago Press, 1938, parags. 22, 35.

[22] Op. cit., 455.

[23] Ibid., 439.

for adequate empirical reference—the latter to be discussed in the following chapter. The parallel serves to indicate that the various systems of logic are a series of hypotheses on the nature of deductive system.

The question of consistency or coherence is, however, not merely a matter for logic: it is, as we already remarked, a prerequisite to verification in empirical science. There is the need for the coherence of the what-answer, the formulation of which can take its place as middle term of the Aristotelian scientific syllogism. But besides the Aristotelian scientific syllogism there is what may be called the syllogism of certitude which relates not explicandum and explanans but hypothesis and evidence: If e, then h; and e; therefore h. Here the focus of attention of the is-question is no longer consistency but fact. The consistency of h is supposed; the grasp of the relation of h to its conditions or to the evidence is expressed in the major premise. In so far as the evidence is known as occurrent, the conclusion follows factually. But the truth value of the conclusion lacks the simplicity of the truth value of an analytic proposition, and that in several ways.

In the first place, the relation expressed in the major premise of h to e, of hypothesis to factual evidence, has a complexity essentially lacking in the analytic proposition, and the connotation of the relation is of the mental stance of the is-question. Again, the evidence available, e,' may fall short of e. In that case there can be grounded probable judgment, and there can be, according to some, a fractional estimate of that groundedness. Finally, the hypothesis h itself is no more than a possibility—to this point we have returned on various occasions already—and on this fundamentally rests the fact that the judgments, the verifications, of science, be they statistical or classical, are no more than probable. The statistical hypothesis, for instance, presupposes a classification of events based on classical science, and revision of that classification through scientific advance leads to the need for the replacement of that hypothesis by a better one.

We may sum up these comments on truth values by saying that truth values refer in general to the acceptability of propositions as analytic or systems as consistent or hypotheses as verified, and that acceptability is such as to involve the answering of an is-question. The truth value 1 signifies a type of unqualified acceptance. Turning now to unit probability we note that it has nothing to do with the acceptability of a hypothesis, proposition or system, nor is it a quality of an is-answer. Unit probability can be a hypothesis for a certain class of events, being verified in so far as the divergence of the frequency of occurrence of those events from unity is random. As any hypothesis, it is the answer to a what-question. It is the peculiar limiting case of the fraction which gives an ideal frequency, and so the non-systematic divergence from it can only occur by the actual frequency falling below it, but otherwise there is nothing unusual about it. The hypothesis, $p = 1$, f-probability, can be verified, and the verification can be more or less probable, v-probability. One might, for instance, raise the statistical what-question, What proportion of laughing bipeds are human? A plausible hypothesis is p = 1. The hypothesis is substantially verified, t = 1, and even in this artificial case one cannot confuse the hypothesis with the quality of the verification in so far as one distinguishes the two types of question, here specifically, What proportion of laughing bipeds are human? and, Is the proportion, p = 1? Part of the difficulty here is that if the answer to the last question is Yes, and the proportion is not merely asymptotically 1, then one may shift one's stance of investigation to that of the classical type and consider as verified the law 'all laughing bipeds are human.' But that shift of stance change's ones attitude towards the 'law,' and the verification is not a matter of counting. The probability of the verification in either the statistical or the classical case is not some fraction that, apart from random differences, corresponds to the relative actual frequency of verifications. A hypothesis does not become highly probable or certain through a preponderance of favourable tests: one contrary test can make it highly improbable. Again, the probability of a verification or a judgment is not reached, as in f-probability, by an insight that abstracts from random differences. Such an insight yields only a hypothesis, an ideal frequency, which requires verification. More contortedly we may argue that the probability of a verification cannot be the ideal frequency of the verification: for then the probability of verifying that ideal frequency would be another ideal frequency, and so *ad infinitum.*

We have been concerned with the structure of the investigation of the scientist. That investigation has its origins in questions about the empirically-given and moves in its simplest form through hypothesis, the answer to a what-question, to verification, the answer to an is-question. This characterizes scientific knowing, and indeed knowing in general. It is basically a triply-structured process, of what is presupposed by questions and prior to answers, of what is reached in so far as what-questions regarding that given are answered, of what is contributed by is-answers which complete the process. Now the crucial issue is, what position to adopt regarding the relation of that structured knowing to reality? In so far as one emphasizes any one element of the structure of knowing at the expense of the other two one may reach a variety of philosophic positions ranging from idealism to materialism. The simplest assumption, however—indeed the most plausible, economic and practical hypothesis, is one which fits uniquely the implicit suppositions of men of common sense and science, even of philosophers. That assumption is the assumption of the isomorphism of the structured knowing with the real. This isomorphism is a far cry from that of logical atomism or that of Spinoza's *ordo idearum est ordo rerum*: it is not an isomorphism of propositional structure and fact or of ideas and fact; it is an isomorphism of the structured anticipation of knowing with the real as its object.[24] On this view the real is what is known or to-be-known in true judgment, or in terms of scientific practice it is

---

[24] Cf. chapter twelve.

what is known or to-be-known by Theory Verified in Instances. This clearly does not make the real a function of what is known in contemporary science and common sense. Reality is neither a Kantian Ideal nor a dependent of the mind. It is there to-be-known and we move asymptotically towards its knowledge through the true judgments of science and common sense. Reality is not a function of what is known: rather, knowing is a function of the real, where the word 'function' has a dual sense. It has the mathematical sense of isomorphism. But it also has the biological sense—and here we might fit our assumption into the context of the evolutionary hypothesis to be discussed later. Knowing is a process which takes place in the human organism. It is, one might say, a higher process of digestion with questions for appetite. It is a process which has evident survival value—witness Western science, the arms race, or merely man against beast. Had that process emerged as Plato or Plotinus described it, we would long since have become extinct. It is its adaptation to the environment, material reality, which ensures its organic possessor survival, and that adaptation is the one we described as an isomorphism.

Is there real randomness? It is the given which leads us in the first place to formulate the hypothesis of randomness, an answer to a what-question. To judge that randomness is objective, real, is to affirm that reality is such that it excludes our systematically understanding it at any stage. Can that conclusion be absolutely certain? No: but it has a peculiar relatively invariant certainty about which we will presently speak.

Again, we might ask the eccentric question, Is the ideal frequency real? No: for, an ideal frequency is an answer to a what-question, and so is only part of the structured knowing or known. What then is real? One can say at least that the real is such that the events in question do not in their occurrence diverge systematically from the verified ideal frequency. One is talking here of a verified structure of events, therefore of what is real in terms of a triply-structured knowing.

Is there real emergence? Again, the question has its origin and formulation within developing science. Its answer will involve the same structured knowing, its discussion will occupy us throughout chapter nine, and there we will come to appreciate more precisely the interrelation of the various components in the structure of knowing. Here too our conclusion will be relatively invariant or invulnerable and we turn immediately to explain what is meant by this.

Earlier we discussed how the results of science were never more than probable, were therefore always falsifiable and open to revision. The peculiarity of the position we have taken on knowing and its relation to reality is that it escapes these restrictions. The position includes elements and structures which are invariant, not open to revision, and also further structures which are so related to these basic elements as to be considered relatively invariant.

The central clue to that basic invariance may be had by noting that science can in general be methodologically specified as Theory Verified in Instances. In other words, the knowledge which advancing science anticipates is a triply-structured knowledge. One might describe this view as a well-established theory of the nature of scientific investigation. At first sight it would appear to be no more invulnerable to falsifiability than any other theory. Yet how is it to be falsified? The process of falsification itself implies an acceptance of the theory, for it would hope to show that the theory was not a theory verified in instances. Moreover, if this view of science is to be adequately replaced it must be by another theory which is verified in instances of scientific practice. And even if one takes the more extreme view that scientific procedure cannot be theorized about, one would nonetheless claim that one's view could be formulated and that it would meet the facts: again implicitly admitting the above methodological specification. Moreover, implicit in the exchange is the assumption of isomorphism. It is the implicit assumption of intelligent and reasonable argument. One seeks to determine what the case is, or to persuade another of what is the case, by argument. Why? Because it is assumed that intelligent and grounded argument is not a flight towards Plotinus' One but a closure on what is the case, on the real. What is involved in all this is what E. Coreth calls 'The Dialectic of Performance and Concept.'[25] In so far as the performer, the metascientist, arrives at a view of science and reality which does not square with his own practice, his view and his practice come into dialectic conflict, a conflict which can ultimately lead to a Revision of his views. The failure of much of philosophy rests on the failure to admit that conflict or to exploit its possibilities. Thus, Hume's attitude as performer was one of questioning and of intelligent criticism of the views of others—and, as our earlier quotation shows,[26] even of his own—but the view that he formulated of human knowing left this performance out of account. Kant was extremely reflective and critical in his judgment of the deficiencies of the current views, but he did not exploit the critical attitude of the is-question in himself when he faced the problem of objectivity, and he did not concede in his theory that his own critical judgments were related to what was the case. Finally, those who consider the present view mistaken do so because they have good reason to do so: but in so far as they formulate their reasons and intelligently criticise the present view they are performers of the view they reject.

It is in this way that the basic conclusions of methodology move out of range, so to speak, of a principle of falsifiability. It remains to note that, according to their degree of proximity to that basis, other methodological conclusions share to that extent in the same invariance. This is a question with which we will deal later in the context of the methodological principle of evolution, where its close dependence on the invariant pattern of classical and statistical scientific methods will be shown to give it a relative invariance. But we might illustrate that type of

---

[25] *Continuum*, 1965, 447–454.

[26] Above, p. 110.

invariance here from another field, from the work of the psychologist Jean Piaget. Piaget has been engaged for over forty years in the study of the development of intelligence in the child, and that study has led him to a complex theory of development in terms of schemes, groups, lattices and groupings, and equilibrium of adaptation. The complex superstructure of the theory may clearly be revised, but underlying that complexity there has become manifest a basic methodological position, a position to some extent paralleled by B. Lonergan's methodological analysis of development.[27]

> Piaget's general "hypothesis" is simply that cognitive development is a coherent process of successive equilibrations of cognitive structures, each structure and its concomitant equilibrium state deriving logically and inevitably from the preceding one. Much of what constitutes his theory is concerned in one way or another with the details of this hypothesis, and it would not be unfair to say that most experiments appear to be set up to demonstrate its validity, rather than to "test" it in any rigorously predictive sense.[28]

Indeed, the basic components in the hypothesis so pertain to the structure of the investigation that one can hardly say that there is question of demonstrating its validity, although it is continually verified. Yet prior to a methodological reflection such hypotheses as this are not understood. Like the system of operations within the child's cognition described by Piaget, they may otherwise scarcely be adverted to.

> The question may arise as to whether the subject himself is aware of the specific structure which his cognitive operations form, or even that they form a structure at all. No, not unless he has been reading Piaget. The system of operations is not itself something upon which the subject can ordinarily focus his cognitive instruments; rather, it is *with* which and *into* which he incorporates the data of the concrete problem before him.[29]

We will return to these questions in the concluding chapter.

---

[27] *Insight*, 451–483; CWL 3, 476–507.

[28] J. H. Flavell, *The Developmental Psychology of Jean Piaget*, New York, Van Nostrand, 1963, 36–37.

[29] Flavell, op. cit., 169.

We will be concerned here with more basic problems of the method of statistics: problems not so much of its axiomatic formulation as of its empirical origin and reference. As M. Loève remarks, 'a stability property is at the root of the whole development of probability theory.'[1] In chapter two we discussed this stability property in terms of the basic insights of probability theory. There we avoided the vexed question of convergence underlying von Mises' axiom of limiting frequencies and followed rather Lonergan's presentation, in which limiting processes are purposely excluded:

> Consider a set of classes of events, P, Q, R ... and suppose that in a sequence of intervals or occasions events in each class occur respectively $p_1, q_1, r_1 ... p_2, q_2, r_2 ... p_i, q_i, r_i ...$ times. Then the sequence of relative actual frequencies of the events will be the series of sets of proper fractions, $p_i/n_i, q_i/n_i, r_i/n_i . . .$, where i = 1, 2, 3 ... and in each case $n_i = p_i + q_i + r_i + ...$ Now if there exists a single set of constant proper fractions, say p/n, q/n, r/n ... such that the differences $p/n - p_i/n_i, q/n - q_i/n_i, r/n - r_i/n_i$ ... are always random, then the constant proper fractions will be the respective probabilities of the classes of events, the association of these probabilities with the classes of events defines a state, and the set of observed relative actual frequencies is a representative sample of the state.[2]

For ordinary experimental procedure this restriction is no great handicap. So, for instance, one can make use of the continuous Normal Law as an evident 'norm' from which actual frequencies diverge randomly without the inclusion of any further complexities regarding convergence of frequencies. Because this is so, the present section may be regarded as an aside upon which the following chapters do not depend. The aside is, however, worthwhile because there are implicit here basic difficulties of a methodological nature, the discussion of which throws light not only on the question in hand but on the general problem of heuristics, of the passage from description to explanation, from nominal to essential definition.

---

[1] Loève, *Probability Theory*, Princeton University Press, 1963, 232. We will take this book as representative of the more general advanced theory of probability and refer to it as 'Loève.' For more detail and application we will use W. Fellar, *Probability Theory and its Application*, New York, Wiley, 1951, Vol. 1, where only discrete sample spaces are considered. As Fellar remarks at the beginning of Vol. 2, New York, 1966, the first volume was both a scientific and pedagogical success, a fact which recommends it for a methodological consideration such as ours. We will refer to Vol. 1 simply as 'Fellar.'

[2] *Insight*, 58; CWL 3, 81; cf. also the concluding lines of 66 in both 1st and 2nd editions (CWL 3, 89), the paragraph beginning "However" in the middle of the page.

From a more mathematical and scientific point of view various authors[3] have noted the manner in which a 'restricted' probability theory would leave unavailable to itself large well-developed areas of mathematics. Again, both Fellar and Loève make evident how problems arise which call for greater refinement. There is, for example, the Petersburg game or paradox[4] in which random variables without finite expectation occur. But even the elementary coin tossing experiment leads us spontaneously to consider an indefinite sequence of experimental results and the question of the peculiarities of the frequency convergence, the elementary Poisson theorem calls for the consideration of a denumerable number of values[5] and provides clues to a discussion of a more general type of convergence,[6] and the de Moivre-Laplace theorem both indicates a line of development in the notion of convergence[7] and shows the need for the consideration of sample spaces involving non-countable sets of values.[8] Our problem here of course is not to solve any mathematical difficulties, but to expose what exactly is involved in these developments, how they are related to the empirical, where there are weaknesses in the discussion. A remark of Hilda Geiringer is relevant here:

> Today we see at one extreme the "consumer of statistics" who wishes to apply ready-made statistics to his problem, be it in medicine, education, or linguistics. ... At the other extreme, stand those mathematicians who are exclusively interested in the mathematical aspect of some problem ... who teach that probability "is" measure theory ... and dismiss frequency theory as "awkward mathematically." Between the extremes, the conception of probability theory as a mathematical science leads to a frequency theory of probability, much in need of the mathematician's ideas and ingenuity, but free of the confusion of task and tool.[9]

In exposing the source of confusion and the possibility of clarification we proceed in our accustomed heuristic manner. One might compare the discussion to follow of the basis of probability theory to W. Sawyer's presentation of abstract algebra.[10] The stress in that presentation is on the pre-formulation stage in learning algebra, on the elements that enable the student himself to cast proofs

---

[3] E.g. Reichenbach, *The Theory of Probability*.

[4] Fellar, 199.

[5] Loève, 15.

[6] Op. cit., 270.

[7] Fellar, cp. 7, section 2.

[8] Loève, 22.

[9] *Mathematical Theory of Probability and Statistics*, 48–9.

[10] *A Concrete Approach to Abstract Algebra*.

into formal expression.[11] Our aim, however, is more fundamental than his. By centering attention on the pre-formulational stage of the theory we hope to pinpoint the sources both of philosophic difficulties and of methodological solutions. As has been regularly noted, these points tend to be passed over in the usual presentations of probability theory with the result that obscurity prevails regarding, e.g., the problem of defining randomness or the role of correspondence rules.

The weakness of the usual frequency theory of probability has been exposed from various points of view. Thus A. H. Copeland Sr.:

> In the frequency theory probabilities are defined as limiting frequencies. We can observe finite frequencies and we believe that they are close to their limits whenever they are constructed out of a large enough number of trials. However, we can only observe finite frequencies and no finite frequency can in any way restrict the limit. Hence differences in limiting frequencies do not correspond to observable differences in behaviour.[12]

Again, Lindsay and Margenau point out, briefly, that after any given number N of coin tossing, long runs of successes or failures can leave the observed frequency differing from the prescribed probability by an amount greater than a preassigned epsilon: worse still, there is the possibility of an unlimited straight run of heads with an unbiased coin.[13] Clearly, here we have difficulties of formulation and verification, and difficulties regarding the interplay of mathematics and physics. These difficulties have been partially met in a variety of ways. Some consider the basic definition satisfactory and add distinctions to expose lacunae in the objections.[14] Others seek to modify the basis of the frequency theory either by allowing only finite probabilities to be included,[15] or by fixing on certain restrictions on the degree of oscillation allowed,[16] or by evolving a clearly finite frequency theory of probability.[17]

---

[11] Cf. Sawyer's remarks, op. cit., 2, 182.

[12] 'A Finite Frequency Theory of Probability,' in *Studies in Mathematics and Mechanics*, presented to Richard von Mises, New York, Academic Press, 1954, 278.

[13] Lindsay and Margenau, *Foundations of Physics*, 165–7.

[14] E.g. von Mises, *Probability, Statistics and Truth*, 86–7; E. Nagel, 'Principles of the Theory of Probability,' *International Encyclopedia of Unified Science*, University of Chicago Press, 1955. Vol. 1, pt. 2, 375.

[15] E.g. Lindsay and Margenau, op. cit., 167.

[16] Hans Blume, *Zeitsch. F. Physik* 92 (1934), 232–52; 94 (1935), 192–203. In *Zentralblatt für Mathem.* 10 (1935), 172, Khintchine remarks that the method employed is nothing more than a circumscription of the concept of limiting value, and does not differ essentially from von Mises' view.

[17] A. H. Copeland, op. cit. footnote 12, 278–284.

The central problem, evidently, is 'to give a precise and acceptable "meaning" to the notion of "clustering of frequencies" which, as we have seen, is at the very root of the interpretation of randomness.'[18] Since meaning is correlative to insight, a precision of meaning will require a fuller basic insight than that already described.[19] A clue to this fuller insight would seem to be offered by Lindsay and Margenau when they remark, in regard to the difficulty of non-convergence discussed above, that

> one supposes instinctively that there must be some correlation between s (the number of tails in a run) and N' (where the run of tails begins) which renders the inequality $[s/(N' + s) > 2\epsilon]$ void. This impression, however, is as false as the supposition that the chances for the occurrence of a rare event increase the longer one waits for it.[20]

The impression would indeed seem to be rendered invalid by the possibility, say, of an uninterrupted series of heads resulting from the tossing of an unbiased coin. Still, both the impression and the undeniable possibility of long sequences of heads or tails offer clues to the broadening of the basis of probability theory through the formulation of 'new types of convergence founded upon notions of measure and unknown in classical analysis.'[21] One appreciates the possibility of indefinitely long runs of either heads or tails, but one has the impression also that such runs are somehow not to be expected. This was not explicitly included when we discussed originally the insight which grasps the stability of actual frequencies in a series of experiments and the randomness of fluctuations from numbers called probabilities. But the broader insight seems relevant both to a development of probability theory and to the exclusion of paradoxes due to more elementary formulations.

But how is the broader insight to be made precise and exactly formulated? First of all it should be noted that the relevant insight is concerned with not just one series of trials but a sequence of series of trials. Inadequately formulated for a particular case, the insight would yield something like the familiar idea that 'almost all sequences of trials would give the same cluster-point value, p.' Behind such an inadequate formulation lurks a collection of difficulties, including such as are touched on by Fellar.[22] Are we discussing limiting processes or not? Are we dealing with infinite or finite sample spaces? These are precisely questions of more adequate formulation. The answer to them will lie within a developed probability theory. Here we are interested, methodologically, in the process towards that developed theory as illustrative of the general process of developing

---

[18] Loève, 14; cf. also 18.
[19] On 123 above.
[20] Op. cit., 166.
[21] Loève, 112.
[22] Fellar, 149–151.

science and mathematics. Our discussion serves to give the lie to Popper's claim that 'there is no such thing as a logical method of having new ideas, or a logical reconstruction of this process.'[23] The process undoubtedly fails to satisfy the canons of deductive logic: but it is certainly not illogical. It occurs in the same basic pattern whether one is advancing from arithmetic to algebra or from classical notions of convergence to definitions of convergence in measure or convergence almost everywhere.

But to return to our problem of formulation. A first attempt at mathematical formulation would perhaps run as follows, with the usual conventions: For $\epsilon >$ 0, $\exists$ N. $|s_n/n - p| < \epsilon$ for $n > N$ for 'almost all' sequences. Clearly, this is vague and mathematically unsatisfactory. Still within classical notions of convergence one might push on to consider, say, m sequences together, so that one has something like:

$$\text{for } \epsilon > 0, n > N, \frac{1}{m} \Sigma_1^m |s_{nm}/n - p| < \epsilon,$$

or one might venture into an ill-defined limit form:

$$\lim_{m \to \infty} \frac{1}{m} \Sigma_1^m \left(\lim_{n \to \infty} s_{nm}/n \right) = p,$$

and call the cluster value, p, the probability.

All this inaccuracy and looseness will undoubtedly cause the mathematician to groan: and this is precisely the point. What is called for is a more accurate formulation, one indeed which will both avail itself of and contribute to the development of convergence theory.[24] The groaning of the mathematician may be concretely illustrated by J. Albertson's criticisms of Lonergan's talk of

absence of further intelligibility in the remainder" as the reason why we go from the relation

$$\frac{1}{2} + \frac{1}{4} + \frac{1}{8} + \ldots + \left(\frac{1}{2}\right)^n = 1 - \left(\frac{1}{2}\right)^n \text{ to the statement}$$

$$\sum_{n=1}^{n=\infty} \left(\frac{1}{2}\right)^n = 1.$$

In reality, however, the transition is justified only by the carefully defined notion of limit and Cauchy sum. The role of definition in the handling of infinite series is even more evident in the case of a series

[23] *Logic of Scientific Discovery*, 32.
[24] cf. Loève, 268ff.

such as $S = 1 - 1 + 1 - 1 + \ldots$, which requires the more advanced concepts of Cesaro or Abel sum.[25]

The point is, and the point seems to have been missed, that prior to the first formulation of the definition of Cauchy or Cesaro or Abel sums there had to be the problem, the initial clues and insights, pointers towards accurate formulation. Consequent on such formulations come applications to various series, applications which can be routine, automatic. The methodological interest, however, is in the pre-formulation stage.

Before discussing further this advance towards precise formulation we must return to the elementary theory of large numbers and to some of its basic insights. Von Mises, in his book *Probability, Statistics and Truth,* devotes a chapter and more[26] to the consideration of the laws of large numbers and to Poisson's use of one of these laws in defining probability. According to von Mises, Poisson, in the early pages of his book[27] gives a definition of the Law of Large Numbers which is in fact basically identical to his, von Mises,' frequency definition of probability. There, Poisson considers this law as the basis of probability theory. Later, however, Poisson takes his start not from this law but from a formal definition of probability introduced by Laplace. Subsequently he deduces, by analytic methods, a mathematical proposition which he also calls the law of Large Numbers.[28] Von Mises shows at some length that this latter Law of Large Numbers is an arithmetic law,[29] and he goes on to point to a certain confusion in Poisson's discussion:

> At the beginning of his (Poisson's) calculations, he meant by the probability ½ of the result "heads" the ratio of the number of favourable cases to that of all equally possible cases. However, he interprets the probability "nearly 1" at the end of the calculation in a different sense. This value was supposed to mean that the corresponding event, the occurrence of between 0.49n and 0.51n heads in a game of n throws, must occur in nearly all games. This change of the meaning of a notion in the course of deduction is obviously not permissible. The exact point where the change takes place is left unspecified.[30]

---

[25] J. Albertson, in a review of *Insight* in *Modern Schoolman* 35 (1957–8), 242.

[26] *Probability, Statistics and Truth,* 21–2 and chapter 4.

[27] *Recherches sur la probabilité des jugements en matière criminelle et en matière civile* (1837); cf. von Mises, op. cit. footnote 26, 21–2

[28] Later we shall see some parallel between Poisson's procedure and that of the textbooks under consideration.

[29] *Probability, Statistics and Truth,* 104–9.

[30] Ibid., 109.

Behind von Mises' discussion lies the general difficulty of the relation of the mathematical to the empirical, but the manifest difficulty is the circularity of Poisson's treatment of probability and the Law of Large Numbers. It is in the solution of the second difficulty that we are immediately interested. Some light will be thrown on the prior general question on the way, and it will be considered more directly later. Later also we will see that our discussion of the theory of large numbers throws light on the equiprobabilist approach to probability.

Our considerations of convergence above had a definite empirical orientation. Even though we allowed ourselves to consider various possible sequences of occurrences, we were throughout interested in understanding actual frequencies of occurrences, indeed we had in mind the particular case of the tossing of an unbiased coin. *What* we wished to understand and explain was fairly clear: in Aquinas's terms we had specified the common matter, a specification which requires nominal definition.[31] A more familiar illustration of this on the mathematical level is the nominal definition of a circle as a uniformly round plane curve, prior to its essential definition in terms of equality of radii. [32] This consideration helps further to show the nature and significance of the axiom of randomness discussed in chapter two:[33] if, for example, von Mises' axiom of randomness is regarded as to some extent determining *what* is to be investigated or understood in probability, then the complaints of lack of clarity of definition levelled against it can be considered to be to a large extent unwarranted. The axiom involves a specification of what is to be understood in an empirical theory of probability, much as one specifies can oval shape' as what is to .be understood by reaching the definition of an ellipse. To ask for a more adequate definition of empirical randomness is somewhat like asking for a more adequate definition of 'oval shape.' What parallels the proper definition of an ellipse is a developed theory of empirical probability, and just as the understanding of an ellipse becomes more refined through the passage from Euclidean methods to Cartesian methods to elliptic function theory, so one might expect refinement in probability theory. Here we would like to stress the possibility of development in probability theory through the distinction of various types of empirical randomness. This development involves the task of pushing forward to a more precise specification of the type of randomness in the empirical object of inquiry, a task to which the various efforts described in chapter two are contributory but not adequate.

But we must return now to our consideration of various properties of large numbers. Here it is clear *what* we wish to understand. The object of our inquiry belongs to the realm of pure mathematics and that object does not involve

---

[31] Cf. B. Lonergan, *Verbum: Word and Idea in Aquinas*, 143–147; CWL 2, 154–58.

[32] Cf. B. Lonergan, *Modern Schoolman* (27) 1949–50, 'A Note on Geometric Possibility,' 125ff. Republished in *Collection*, London and New York, Herder and Herder, 1967; CWL 4, 96ff.

[33] Cf. above, p. 23.

randomness. Again our interest is pre-formulational. Euclid's proof of the nonexistence of a greatest prime may be initiated by noting that[34]

the number $1 + 2$ is not divisible by 2;
the number $1 + 2.3$ is not divisible by 2 or 3;
the number $1 + 2.3.5$ is not divisible by 2 or 3 or 5;

. . . .

The suggestion ends with dots, connoting that one should have caught on. Similarly the following suggestion conveys an initial insight into the property of binary numbers of n digits relevant to a large number theorem:

there is only one number without zeros: 1111 ... 1,
there are n numbers with one zero: 01111 ... , 101 ... , etc.
there are $n(n - 1)/2$ numbers with 2 zeros: 00111 ... , 010 ... , etc., etc.

The various etceteras connote that there are points to be grasped: roughly, one comes to grasp that there is a 'clustering' of numbers as the number of zeros mounts to equality with the number of units, and that this clustering is more pronounced the greater n is. Our hints however are very inadequate: there is indeed a complex of elementary insights involved, insights into the irrelevance of order, insights into the significance of identity of type etc.—the insights embedded in the ordinary theory of permutations and combinations.[35] But at any rate there is little difficulty in appreciating the formula $n!/k!(n-k)!$ for the number of numbers with k similar of one kind, n−k similar and of the other kind. Again one can easily switch to the proportion of such n-digit numbers involved by multiplication by $(\frac{1}{2})^n : (n!/k!(n-k)\,!)\,(\frac{1}{2})^n$. We have spelled out the obvious here, cautious of our presuppositions, in order to emphasize the contrast between this derivation of what in fact corresponds in structure to a particular case of the binomial distribution and the derivation of Fellar.[36] Fellar arrives at the formula $b\,(k; n, p) = \binom{n}{k}p^k q^{n-k}$, where b (k;n,p) is the probability that n Bernoulli trials with probability p for success and $q = 1 - p$ for failure results in k successes and n − k failures $(0 \leq k \leq n)$. The formula is derived in the context of probability theory and the reader spontaneously associates the fractions p and q with the notion of probability introduced earlier.[37] Our own formula is a particular case of Fellar's formula: $b\left(k; n, \frac{1}{2}\right) = \binom{n}{k}\left(\frac{1}{2}\right)^k\left(\frac{1}{2}\right)^{n-k}$ but the reader was not encouraged at any stage to think of the fraction $\frac{1}{2}$ as a probability. There is a switch here which may seem unimportant but underlying that switch of attention or meaning lies

---

[34] Cf. E. O'Connor, *Continuum* 2 (1964), in the Introduction, 315.

[35] E.g. one might consider chapter 2 of Fellar, distinguishing the insights which he formulates from those which remain 'casual.'

[36] Fellar, 105–6.

[37] Op. cit., Introduction and 18–9.

the problem of providing a definite correspondence rule and of making explicit the link between the mathematical and the empirical. When the link is not clearly defined in probability theory, the discussion can oscillate indefinitely between some form of measure theory and experimental considerations.

Through our simple example we have perhaps gone some way towards pinpointing the link, the shift of meaning in moving between number theory and empirical probability theory. In its more elaborate form the conclusion from the theory of numbers is that the proportion of numbers with equal numbers of zeros and units tends towards unity, with increasing n, in a well-defined way. There is no question of randomness being involved. It is a conclusion, following from the elementary properties of numbers already noted, and so if one asks why the convergence is such as it is, the reasons to be given are the premises of the proof. On the other hand, there is the empirical convergence of a less well-defined type discussed earlier, but this ill-defined convergence is not the conclusion of a theorem. Here, if one asks why there is such convergence, the answer does in fact involve mathematical theory but the fundamental answer to the 'why?' is one proper to empirical probability. It is the answer considered already in earlier sections: [38] there is such convergence because concrete non-systematic occurrence excludes a certain type of systematic divergence.[39] However, as just remarked, mathematical theory is called in because, as well as the methodological reason for convergence there is the form of the convergence to be discussed, and a form of mathematical convergence akin to that involved in Large Number Laws is seen to fit the bill to some extent. Our discussion serves to focus the limitations in this kinship and to make evident the need for mathematical ingenuity in developing an adequate empirical theory. Equiprobabilism would no doubt lay emphasis on the generalizations and developments possible on the mathematical side: thus one may pass from the binary sequence we used as illustration to n-ary sequences, where n is a common denominator for a set of probabilities, and by the process called 'mixing' by von Mises, reach a plausible basis for *a priori* discussion of any set of rational probabilities. Obviously this is not invalid: such notions indeed find more adequate expression within a theory of random variables, and there one may arrive at the view that e.g., 'height is a cumulative effect of many such random variables.'[40] The failure lies in the inadequate empirical reference and dependence. A more genuine development requires some centering of attention on problems of empirical convergence, without which, for example, adequate operative distinctions between what is mathematically possible and what is empirically probable cannot emerge. In the beginning of our discussion we emphasized this attitude by touching on the clues and insights

---

[38] Cf. above, p. 124 ff.
[39] Cf. Lonergan, *Insight*, 103–105; CWL 3, 126–28.
[40] Fellar, 204.

which in fact lead to technical definitions of convergence in measure and converge almost everywhere. Within that initial discussion was also the suggestion that empirical probability requires definition within the context of a more refined consideration of convergence.

It is to be noted that such a way of introducing a definition of probability would differ from the normal presentation. Usually a preliminary notion of probability is developed and used within the theory to gradually define various types of convergence. Thus, Fellar remarks at the beginning of his book that 'the extent to which logic, intuition, and physical experience are interdependent is a difficult problem of philosophy into which we need not enter.'[41] In fact, however, absence of clarity on this issue tends to leave obscurity in the subsequent treatment, where one is appealing to mathematics, where to experience, even indeed what the entire subject is about. So, Fellar remarks that 'we shall no more attempt to explain the "true meaning" of probability than the modern physicist dwells on the real meaning of mass and energy or the geometer explains the nature of a point.'[42] Such a view seems to bow to some extent to the myth that metaphysics or philosophy reaches the 'true meaning' in these cases. On our view, the meaning of probability is precisely what a treatise on probability is after. That meaning can be virtually contained in an axiom system, and the meaning is mathematical only or also empirical depending on whether or not suitable correspondence rules are included or understood. To this we will return later. To come to Fellar's actual development of the theory: his introduction of 'probability' is based on the view that 'the intuitive meaning of probability is clear, but only as the theory proceeds shall we be able to see how it is applied,'[43] and his illustrations are the usual examples of coin tossing, dice throwing etc. At the end of the introductory discussion he puts forward the 'fundamental convention':

> Given a discrete sample space S with sample points E_1, E_2, ... , we shall assume that with each point E_i there is associated a number, called the probability of E_i and denoted by $\Pr(E_i)$. It is non-negative and such that $\Pr(E_1)+\Pr(E_2)+ ... = 1$.[44]

The convention clearly serves as a simplified introduction of normed additive measure theory, but the linkup with the previous empirical illustrations is not clearly made and, as we noted above,[45] the empirical illustrations give rise to a variety of fundamental problems. Only in chapter seven does Fellar return to the task of more adequate formulation:

---

[41] Op. cit., 1–2.
[42] Op. cit., 3.
[43] Op. cit., 7.
[44] Op. cit., 18–9.
[45] Cf. above, p. 150.

On several occasions we have mentioned that our *intuitive notion of probability* is based on the following assumption. If in *n* identical trials A occurs *v* times, and if *n* is very large, then *v/n* should be near the probability p of A. Clearly, a formal mathematical theory can never refer directly to real life, but it should at least provide theoretical counterparts to the phenomenon which it tries to explain. Accordingly, we require that the vague introductory remark be made precise in the form of a theorem.[46]

The theorem, or rather theorems, in question are the weak and strong laws of large numbers,[47] proved in various parts of chapters seven, eight and ten. The relation of the 'intuitive notion of probability' to the whole process is not made clear and so to some extent Fellar's treatment would merit a criticism analogous to von Mises' criticism of Poisson's discussion. Fellar considers these theorems to formulate in some way the intuitive notion of probability:

> Taken in conjunction with our theorem on the impossibility of gambling systems, the law of large numbers implies the existence of the limit $(s_n/n \rightarrow p)$ not only for the original sequence of trials but also for all subsequences obtained in accordance with the rules of section 2. *Thus the two theorems together describe the fundamental properties of randomness which are inherent in the intuitive notion of probability and whose importance was stressed with special emphasis by von Mises.*[48]

One might question whether the two theorems are an adequate description of randomness, and whether the properties of randomness are inherent in the intuitive notion as presented. Again, one would like to see a more formal presentation of the theorem on the impossibility of gambling systems which would reveal its precise relation to the intuitive notion of probability. These points of criticism are of course methodological rather than mathematical: they serve to expose and illustrate the inescapable obscurities which result from the basic difficulties which Fellar mentions at the beginning of his book.

Loève's discussion of probability, which is not restricted, as Fellar's was, to denumerable sample spaces, gives evidence of similar obscurities. In his introductory part[49] he seeks

> to give "intuitive meaning" to the concepts and problems of probability theory. First by analysing some ideas derived from everyday experience—especially from games of chance—we shall

---

[46] Fellar, 141; italics his.

[47] The weak law can be considered as equivalent to convergence in measure, the strong law as equivalent to convergence almost everywhere.

[48] Fellar, 157; italics his.

[49] Loève, 1–51.

arrive at an elementary axiomatic setup ... Then we shall apply this axiomatic setup to describe in a precise manner and to investigate in a rigorous fashion a few of the "intuitive notions" relative to randomness.[50]

One might well question in detail the methodological presuppositions of this approach. Basically the approach implicitly assumes a conceptualist account of knowing which is in radical opposition to our own view, and it suffers from a consequent vagueness about the nature of 'intuition,' or insight, and about the origin of ideas. The most obvious contrast with our own position is the inversion of the order of concept and insight. As we have continually illustrated, the concept, the formulation, is posterior to the insight.[51] If this is ignored, the mathematics may not be affected, but foundation questions will,[52] and of course questions of pedagogy.

Loève introduces the notion of probability quite simply:

Let the *frequency* of an outcome A in *n* repeated trials be the ratio n_A/n of the number n_A of occurrences of A to the total number *n* of trials. If, in repeating a trial a large number of times, the observed frequencies of any one of its outcomes A cluster about some number the trial is then said to be *random* ... The outcomes of a random trial are called *random* (chance) *events*. The number measured by the observed frequencies of a random event A is called the probability of A and is denoted by PA.[53]

Later he acknowledges that 'we ought to give a precise and acceptable "meaning" to the notion of "clustering of frequencies" which, as we have seen, is at the very root of the interpretation of randomness,'[54] but, as with Fellar, the relation of his formulation of a precise notion, through the Bernoulli and Borel laws of large numbers, to the prior intuitive notion, is not clear. Twice in this section he touches precisely on the link between mathematics and experiment, each time using the phrase 'in other words,' and in each case the phrase is misleading. It is worthwhile quoting the two passages:

---

[50] Op. cit., 1–2.

[51] A series of Penguin Books on mathematics, elementary and advanced, by W. W. Sawyer, abundantly illustrate the point. Gf. also footnote 10 above. Also E. D. Hutchinson, *A Study in Interpersonal Relations*, ed. P. Mullahy, New York, Hermitage, 1950, 'The period of Frustration in Creative Endeavour,' 404–420, and other essays there; J. Hadamard, *The Psychology of Invention in the Mathematical Field*, New York, 1945.

[52] Cf. P. McShane 'The Foundations of Mathematics.' *Modern Schoolman*, 40 (1962–3), 373–387.

[53] Loève, 5; italics his. It is interesting to note that Loève belongs to the group who consider probability to be a property.

[54] Op. cit., 14; cf. also 18.

Bernoulli Law of Large Numbers (1713). In the Bernoulli case, for every $\epsilon > 0$, as $n \to \infty$, $P(|s_n/n - p| > \epsilon) \to 0$. In other words, the probability distribution of the frequency s_n/n of an outcome in n repeated trials concentrates at the value p of the probability of the outcome, as the number of trials increases indefinitely.[55]

We organise repeated independent trials where we observe the value of X; in other words, we consider independent random variables X_1, X_2, ... with the same probability distribution as X.[56]

To go into the theory of random variables etc. here would take us too far afield, but it should be clear from the previous discussion of empirical convergence, and mathematical laws of large numbers and convergence, and also from von Mises' criticism of Poisson referred to, that behind the phrase 'in other words' there is a shifting from mathematical to empirical. This is the sort of shifting that would call for the formulation of correspondence rules. We do not wish to exaggerate the importance of such correspondence rules. Indeed, in normal practice one may apply to this question the following comment:

> The real principle of solution is neither one rule nor any set of rules but rather the fashioner of all rules, intelligence itself in act, determining what it takes as relevant to itself and so *de ratione speciei* and what it dismisses as irrelevant to itself and so as pertaining to the *partes materiae*.[57]

Nevertheless, in present probability theory there are difficulties in this region which require more delicate and formal treatment. In a certain sense there is a gap to be located, acknowledged, gradually and asymptotically filled. Nor is this merely a matter of correspondence rules, for it involves the basic notion of probability which is the pivot of development in the theory. According to Fellar,

> the truth is that, like all mathematics, the theory of probability builds theoretical models which are applied in many and variable ways. The technique of applications can be understood only after the theory. The intuition develops with the theory. Our statistical description of probability suffices as an intuitive background for the beginning. It is vague, but, to use a simile of T. E. Lawrence, it is "like the bow of the Mauretania: the bow has so much weight behind it that it does not need to be as sharp as a razor."[58]

---

[55] Op. cit., 14.
[56] Op. cit., 20.
[57] B. Lonergan, *Verbum: Word and Idea in Aquinas*, 146; CWL 2, 157.
[58] Fellar, 5–6.

There are various weaknesses in this view. The theory of probability is not just 'like all mathematics': it is more closely concerned with the concrete than classical mathematics. Unless the techniques of application play a precise role in the development of the theory, the theory is liable to generate pseudo-techniques of application which diverge from empirical requirements. Here of course we are thinking mainly of 'ordinary' probability theory, but there are undoubtedly parallels in quantum statistics. The initial 'statistical description' is not merely 'an intuitive background for the beginning,' it runs through the entire structure. Therefore, far from being vague but satisfactory, it needs to be 'as sharp as a razor.' Or, switching to Lonergan's metaphor of the scissors movement of empirical inquiry with an upper mathematical blade and a lower empirical blade,[59] one must appreciate which blade is being developed or in need of development, and one must develop harmoniously the two blades, if the scissors movement is to operate successfully. One needs, then, to pinpoint precisely where the development is to occur, in what sense 'the intuition develops with the theory.' Will the intuition develop, for example, in the manner in which the basic insights of Lattice Theory can be developed, by refinement and addition,[60] or will it develop in a somewhat equivocal sense, for example, as Newton's intuition regarding uniform motion was transformed by Einstein's?

Certainly it should be clear by now that the basic gap is one between the possible, in a mathematical sense, and the probable in an empirical sense. Our basic example of coin tossing, in the discussion of Large Number Laws, perhaps best illustrated the point. It is very evident that a run of n heads, considered from a mathematical point of views, is a possibility no matter how large n: but there is quite a shift of meaning involved in moving to the question of the empirical probability of a run of n heads, or of the falsifiability of any such probability.

Again, Large Number Laws do not of themselves close the gap. The bridging of the gap must involve intrinsically the basic notion of probability. One may see this in a new and more generally fruitful light by considering that in the case of large number theorems the convergence involved is conditionally necessary, while in the case of empirical inquiry, the convergence involved is a *de facto* affair. The convergence in the case of large number theorems is necessary: granted the premises and conditions—and there is quite a range of these[61]—we can strictly deduce that convergence. The empirical convergence, on the other hand, is a given which we seek to understand, and the understanding is expressed, not in terms of necessity, but in terms of possibility. Another enlightening parallel to this situation is had by comparing the principle of contradiction as Aristotle conceived it to the parallel postulate of Euclidean geometry. The former is

---

[59] Lonergan, *Insight*, 312–4; CWL 3, 337–38.

[60] Cf. G. Birkoff, *Lattice Theory*, New York, Amer. Math. Soc. Colloq. Publ., 1948.

[61] Cf., e.g., Loève, ch. VI.

grasped as a necessity of thought and of reality; the latter is grasped not as a necessity but as a possibility, a possibility which is in fact the intelligible correlative to Euclid's nominal definition of a plane surface, 'a surface which lies evenly with the straight lines in itself.'[62]

> In the light of our analysis it seems reasonable to say that the fifth postulate stands to the nominally defined plane surface as does formal cause to common matter. A formal cause is an intelligible correlation or unification. Common matter is the correlated. Combine the two and there result necessary properties which, in the present instance, are the properties of the Euclidean plane surface. As the equality of radii stands to the circle, as the equality of right angles stands to the straight line, so the parallel postulate stands to the Euclidean plane surface.[63]

As the parallel postulate is to the nominal definition of the plane surface, so will any attempt to formulate an intelligibility of empirical probability be to its common matter. The common matter in the latter case is such as to elude neat specification, as our earlier discussion of randomness showed.[64] One may recall here P. Boutroux's discussion of the scientific ideal of the mathematician.[65] In contrast with earlier ideals, where aesthetic or formal limitations were current, the modern ideal consists in specifying the mathematical object—say, an ellipse— and endeavouring to exhaust its intelligibility through a series of ever more adequate techniques e.g., Euclidean, Cartesian, elliptic function, etc., techniques. But in the case of probability it is not merely a matter of circling round a mathematical object. The question is empirical and the object is, one may say, more elusive. It is not then a matter of working out mathematically a series of manageable definitions of randomness etc.—though this too is involved—but of taking into account the contrariness of empirical randomness in its various types, and this in a way which will modify intrinsically the initial inadequate notion and formulation of probability. It was to open up this viewpoint that we presented initially, prior to the discussion of any author, the peculiar hints regarding the solution of difficulties associated with the elementary definition of probability as, say $\lim_{n\to\infty} \frac{s_n}{n}$. Our basic point throughout is that advance in probability theory requires the transformation of the central insight according to the particular conditions of randomness to which the theory is relevant. Our suggestion, scarcely more than an illustration, nonetheless did in fact point in a direction of

---

[62] Charles Thomas-Stanford, *Early Editions of Euclid's Elements*, London, 1926, 153.

[63] Lonergan, 'A Note on Geometric Possibility,' *Modern Schoolman* (27)1949–50, 132. Republished in *Collection*, London and New York, Herder and Herder, 1967, 105; CWL 4, 100.

[64] The discussion of this in chapter 2 was more from the viewpoint of formal specification; the present chapter concentrated more on empirical specification.

[65] P. Boutroux, *L'dèal Scientifique des Mathematiciens*, Paris, 1955.

genuine development. So, for instance, even the uninitiated may appreciate that a series of uninterrupted heads, or tails, which is a mathematical possibility can, so to speak, be stored in the limbo of the physically non-probable by linking zero measure with zero probability: the series in question is a legitimate sample point, but lost in a continuum of sample points.

There is nothing novel in this idea, nor is it adequate to resolving all the difficulties already mentioned. It is closely related, in particular, to the strong Law of Large Numbers and to Tchebichev's inequality,[66] and in general to the range of convergence theorems to which a large section of Loève's book is devoted.[67] What seems deficient is the general orientation of the investigation. Certainly, such a development of the 'mathematical blade' is required: but that the explicit definition of random variables and the discussion of various types of convergence and conditions can remain too safely mathematical could be more clearly adverted to, and the initial intuitive notion of probability could be more explicitly modified and integrated into the theory. One may note something of this, let us say, inevitable mathematician's stress, in Loève's discussion of the relations between measure theory and probability theory:

> Measure theory investigates families of functions on a measure space to other spaces, distinct or not from the first. On the other hand probability theory has developed and continues to develop intuitions, problems, and methods of its own in exploring those properties of families of functions which remain invariant under all the transformations which preserve their joint distributions—the reason being that the primary datum in random phenomena is not the probability space but the joint distributions of the families of random variables which describe the characteristics of the phenomena.[68]

The parallel relation which he describes in order to aid in an understanding of this is that between geometry and analysis. Geometry remains a distinct science not because it is older than algebra or analysis—just as probability theory is older than measure theory—nor because it has its own terminology, but because it studies properties of sets of points that remain invariant under certain transformations, like Euclidean displacements. The restriction to certain transformations here is analogous to the restriction to distribution-preserving transformations in probability theory. The distinction, then, between probability theory and measure theory does not lie, according to Loève, 'in a greater or lesser generality of the concepts but in the properties investigated in these branches of mathematics.'[69]

---

[66] Cf. Loève, index under Tchebichev, or Fellar, index under Chebyshev.
[67] Loève, ch. VI, 'Central Limit Problem,' 268–334.
[68] Op. cit., 171.
[69] Op. cit., 170.

The concluding phrase of the last quotation pinpoints the tendency. The crux is that the fundamental notion of probability may not be developed in a manner intrinsic to the theory and the pressure from the empirical for more adequate definition and more relevant developments may not be sufficiently felt. It is to this type of remoteness from the concrete that both von Mises and H. Geiringer object. Speaking of Kolmogorov's work on probability, H. Geiringer remarks that 'in contrast to Tornier and von Mises, his basic field is of necessity indeterminate.'[70] The reason for this indeterminacy lies in fact in the power of implicit definition, which is evident in the axioms of Kolmogorov as it is, say, in Hilbert's axioms involving 'point' and 'line,' or in the axioms of Lattice Theory. Stefan Banach's introductory remarks to his paper 'Sur les opération dans les ensembles abstraits et leur application aux équations intégrals,' where the basic notions of Banach spaces were developed, puts the matter well:

> This present work has the object of establishing certain theorems that hold in several different branches of mathematics, which will be specified later. However, in order to avoid proving these theorems for each branch individually, which would be very wearisome, I have chosen a different way, which is this: I consider in a general way sets of elements for which I postulate certain properties. From these I deduce theorems and then I prove for each separate branch of mathematics that the postulates adopted are true.[71]

The power of implicit definition is that it allows a move to greater generality by prescinding from the matter, from nominal definition, so that all isomorphic cases are included. It is a stage in abstraction which allows one to deal purely with relations, terms and relations being left in this sense indeterminate. Obviously the procedure is of value. But since it deals only with what is common to all the cases it is free from the pressures towards relevant development in the particular cases: in so far as nominal definition is called upon, it is a matter merely of application or exemplification.

One could compare the position of probability theory with its elementary form of the 'intuitive notion of probability' with the position of Euclidean geometry prior to the development of the series of non-Euclidean geometries, and in that sense one might expect the development of a variety of probability theories with different degrees and regions of empirical reference.

We have not concerned ourselves here with the axiomatization of probability theory.[72] Our interest was, not the logic of formulation, but the difficulty and the procedure of moving towards more adequate formulation in the

---

[70] Op. cit., footnote 9.

[71] *Fundamenta Mathematica* (3) 1922, 134. The translation is by W. Sawyer, *A Path to Modern Mathematics*, Gretna, LA, Pelican, 1966, 193.

[72] Cf. P. Suppes, *Introduction to Logic*, 274–291, and the references there.

particular case of empirical probability theory. The discussion in turn throws light on the more general problem of the logic of discovery.

If the laws of behaviour enter into the physical system at any point they must constitute either primitive laws in that system or be deducible from the primitive laws. There is no other way out of it. These supposed emergent laws are either epiphenomena or physical laws. For if they represent irreconcilable inconsistencies in the physical system, they are not laws at all, but statements of chance occurrences.[1]

At first sight the problem of emergent laws or qualities would seem to have little to do with our present problem of real randomness. It is the purpose of this section to show that in fact the occurrence of real randomness is intimately connected with the possibility of emergent laws: that, more precisely, real randomness is a necessary and, in a certain sense, a sufficient condition for the occurrence of emergent laws. This conclusion will serve first of all to strengthen the grounds for affirming the occurrence of real randomness: for, if there are emergent laws, in a sense to be defined—and this is an empirical question—then necessarily there occurs real randomness. The conclusion will also throw light on certain aspects of evolution theory, and on the possibility of independent sciences or levels of inquiry. Finally, the discussion will lead to a more precise determination of the meaning of randomness on various levels of empirical science and so also of the manner in which it can be subjected to system either by the correlations of another level of science or by statistics.

The two disputes regarding the possibility of emergent laws and the existence of irreducible sciences may be regarded respectively as the temporal and nontemporal aspects of a single problem, for as G. Bergmann remarked, 'essentially novelty of laws has as little to do with time as the process of inference with the temporal unfolding of the natural processes which such inference might describe.'[2] That single problem is the one expressed by S. Pepper in the quotation with which we began. His paper of forty years ago has recently been revisited by Meehl and Sellars, whose aim was 'not to defend an emergentist picture of this world, but rather to criticize an argument which, if successful, would make the picture indefensible.'[3] Here we hope to expose both the weaknesses of the Meehl and Sellars criticism and the flaws in Pepper's argument. Positively we hope to show that emergent laws do not 'represent irreconcilable inconsistencies in the physical system,' that they are not 'statements of chance occurrences' but

---

[1] Stephen C. Pepper, 'Emergence,' *Journal of Philosophy* (23), 1926, 244.

[2] G. Bergmann, 'Holism, Historicism and Emergence,' *Philosophy of Science* (11), 1944, 213.

[3] P. E. Meehl and Wilfrid Sellars, 'The Concept of Emergence,' *Minnesota Studies in the Philosophy of Science*, Minneapolis, 1962, Vol. 1, 239.

irreducible higher systematizations of aggregates of random but determinate lower events.

Various authors have proposed different explanations of, and criteria for, emergence. For the moment we may take Pepper's characterization as sufficient:

> Emergence ... is a cumulative change, a change in which certain characteristics supervene upon other characteristics ... The theory of emergence involves three propositions: (1) That there are levels of existence defined in terms of degrees of integration; (2) That there are marks which distinguish these levels from one another over and above the degrees of integration; (3) That it is impossible to deduce the marks of a higher level from those of a lower level, and perhaps also (though this is not clear) impossible to deduce marks of a lower from those of a higher.[4]

Pepper sets out to show that the second and third of these propositions are subject to dilemmas: (1) either the alleged emergent change is not cumulative or it is epiphenomenal; (2) either the alleged emergent change is predictable like any physical change, or it is epiphenomenal.

First of all we should ask what precisely is emergent: is it a quality, a law, a characteristic—or is there anything more than a verbal distinction between these? Meehl and Sellars draw attention to the 'philosophical virtuosity of the term "characteristic,"'[5] and Pepper himself draws attention to the ambiguity of the term 'quality,' which frequently covers both quality in Alexander's sense and law of activity. He considers that what the emergent evolutionist has in mind are emergent laws. Before going on it would be well to have further clarification of this issue. Pepper's own remarks give the line of such clarification: 'Accurately speaking, we must first observe, laws cannot emerge. Emergence is supposed to be a cosmic affair and laws are descriptions. What emerges are not laws, but what the laws describe.'[6] Still, what the laws describe must in some sense be described prior to the formulation of laws, and to remove the ambiguity implied in this last remark we may perhaps use the terminological device of saying that the laws *explain* what is pre-scientifically described. Moreover, this distinction between explanation and description would seem to be more than a terminological convenience, for the transition from prescientific description to scientific explanation involves the discontinuity of a switch from relatedness of things to us to correlation of things among themselves. It is the discontinuous switch from heat as felt to heat as defined by the correlations implicit in thermodynamic equations, or from colours as seen to wave and photon theories. The relation

---

4 Pepper, op. cit., 241, Meehl and Sellars (M. and S. in future references), 240.
5 M. and S., 244.
6 Pepper, op. cit., 242.

between the two brings up the problem of correspondence rules, but at all events it is clear that without the prescientific description there can be neither transition nor correspondence rules. It is this which Broad has in mind when he remarks that 'we must wait till we meet with an actual instance of an object of higher order'[7] before we can come to higher laws. It is this too which underlies Pap's discussion of emergence and his conclusion that

> whether a quality or law is emergent does not depend on the stage of scientific knowledge but rather on the question whether certain predicates are only ostensively definable. Specifically, a law correlating a quality Q with causal conditions of its occurrence can, without obscurantism, be argued to be *a priori unpredictable* if the predicate designating Q is only ostensively definable.[8]

Finally, failure to note that the same distinction between description and explanation should be operative in discussions of space, extension and motion leads either to the error of the Aristotelians or to that of Galileo. The scholastic Aristotelians clung for centuries to the description '*partes extra partes*' as if it defined extension; Galileo failed to note that extension, like colour, can be both described and explained. The correct distinction is not that between primary and secondary qualities but that between explanation and description. One might, then, very well clear up the ambiguity of the term 'quality' by associating quality with description, and law with explanation.

Let us now return to Pepper's central argument and the criticism of it by Meehl and Sellars. Briefly, Pepper would claim that if a function f_1 (q, r, s, t) adequately described the interrelationships of the four variables q, r, s, t on a given level, then there can be no other adequate description which is not identical with f_1. If f_2 (q, r, s, t) were considered to describe the interrelationships 'after the integration' or putative emergence, then since sheer time difference should be of no consequence f_2 must be an adequate description 'before' the integration. The point is, either f_1 adequately describes the interrelationships of q, r, s, t or f_2 does;

---

[7] C. D. Broad, *The Mind and its Place in Nature*, New York, Harcourt Brace, 1929, 79. C. Lloyd Morgan's *Emergent Evolution*, London, Williams and Norgate, 1923, appeared six years before Broad's book and in it he traces back the discussion of these questions (p. 2): 'The concept of emergence was dealt with (to go no further back) by J. S. Mill in his Logic (Bk. II I, ch. VI, 2) under the discussion of "heteropathic laws" . . . the word "emergent" as contrasted with "resultant" was suggested by G. H. Lewes in his *Problems of Life and Mind* (Vol. II, Prob. V, ch. iii, 412).' More particularly, he reveals Broad to be echoing Lewes' view (p. 5 of 'Emergent Evolution'): 'Lewes says that the nature of emergent characters can only be learnt by experience of their occurrence.'

[8] A. Pap, 'The Concept of Absolute Emergence,' *British Journal for the Philosophy of Science* (2), 1951–52, 304. This article is reproduced in his *Introduction to the Philosophy of Science*, 364–372.

or if neither adequately describes the interrelationships there is some f_3 that does, but there cannot be two adequate descriptions of the same interrelationships among the same variables.'[9]

An emergent law must, therefore, involve the emergence of new variables. But these variables either have some functional relationship with the rest of the lower level variables or they have not. If they have not, they are sheer epiphenomena, and the view resolves itself into a theory of qualitative emergence, without laws. If they have, they have to be included among the total set of variables described by the lower level functional relations; they have to drop down and take their places among the lower level variables as elements in a lower level functional relation of change.

'Such being the case, our dilemma is established as far as concerns cumulative change—either there is no such thing or it is epiphenomenal.'[10]

Meehl and Sellars criticize the first part of Pepper's argument as being too strong. According to them, what the emergentist says is that there is a region in the four space q, r, s, t, the lower level of integration, within which $f_1 (q, r, s, t) = 0$: e.g., physico-chemical processes which are not occurring in protoplasm. But there is another region, the emergent region, where $f_2 = 0$ holds, $f_1 \neq f_2$.

That such a description fits the orthodox emergentist's view is extremely doubtful, and Meehl and Sellars themselves acknowledge that it hardly merits the name 'emergence': for the description involves no notion of 'supervenience' whatsoever. They go on to acknowledge that most emergentists have something more in mind—the emergence of 'properties.' Nonetheless they stand by their initial criticism and it dominates the remainder of their discussion. Let us briefly summarize their conclusions before undertaking a more detailed criticism of the two positions and a more adequate defence of the emergentist view.

Meehl and Sellars now turn their attention to the second half of Pepper's argument as given in the quotation above. They tend to concede his point: 'If determinism is assumed, so that these variables are themselves lawfully related to the lower level variables, then it must be granted that descriptive laws predicting the course of the latter can, in principle, always be formulated in terms of them alone.'[11] If this is the case, then the function which would adequately describe the interrelationships—call it $E (q, r, s, t, a, b)$ where a, b are the emergents—can be written without a and b, e.g., as $E (q, r, s, t, g (q, r), h (s, t))$ or $f_3 (q, r, s, t)$. This should lead at once to Pepper's conclusion, but, according to Meehl and Sellars,

> once we drop, as we have seen we must, the assumption that f_3 and f_1
> are intended by the emergentist to hold for the same region in $(q, r, s,$
> $t)$ space (f_3 presumably holds for all regions, f_1 only for the "lower level

---

[9] Pepper, op. cit., 243.

[10] Ibid., 244.

[11] M. and S., 247.

of integration") the argument falls apart. For while the emergentist must indeed admit that if f_3 and f_1 are equivalent, then a and b make no difference, it is open to him to say that the difference made by a and b is just the fact that f_1 (q, r, s, t), which holds in regions of q, r, s, t—space which are unaccompanied by a and b, is not equivalent to the function which holds of these variables for regions in which they are accompanied by a and/or b.[12]

The a's and b's are, say, 'raw feels' which do not occur in the presence of matter generally, but only in the presence of matter as it is in the living brain. But here the authors are forced to acknowledge a central difficulty:

> While it is true that prior to the examination of living brains, the function f_1 was quite adequate, and though afterwards we say that f_2 was required for the case of brains, this does not force us to introduce the new variables a and b. For, as we have seen, a and b are eliminable from the descriptive laws.[13]

Here the authors might well have fallen back on their own previous suggestion, which parallels Broad's view on the explanation of secondary qualities referred to above:[14] 'But how will the scientist be led to introduce raw feels into his picture of the world? ... one answer would be that, after all, we experience raw feels, and it is the business of science to fit them into its world view.'[15] However, at this stage in the argument they suggest that 'the introduction of these new variables might be "forced" upon us by theoretical necessities.'[16] How this would occur is not made clear by the authors: they simply claim that an adequate theory, making possible the derivation of the more general function obeyed say within the brain, will include other terms than those which were sufficient for an explanation of pre-emergent phenomena, terms to which the variables a, b pertain, and that it should not be supposed that these new terms are analyzable into the entities for which q, r, s, t suffice as descriptive functors.

> If this seems odd, one should remember that whenever a theory is "correct" it means that we have succeeded in formulating a lawful relation between a value, x, appertaining to the theoretical entity and a value, y, taken on by the observed. Hence in the present case we can write an equation explicitly relating these values, $x = f(y)$. But the fact that we can write this equation obviously does not mean that the entity to which the value x appertains is being equated with the situation to

---

[12] Op. cit., 248.
[13] Op. cit., 250.
[14] Page 143.
[15] M. and S., 250.
[16] Ibid.

which the value y appertains, any more than the discovery of a functional relation between a person's height and weight would require us to suppose that somehow a person's height is the same thing as his weight. Now an argument offered by Pepper in the closing section of his paper hinges partly on a failure to make this distinction.[17]

Only later will we be in a position to expose the weaknesses of the position expressed in this quotation. At any rate, Meehl and Sellars would claim that this analysis shows that the emergent terms or laws cannot be absorbed, as Pepper would have it, into the lower-level system, even though the equations $a = g\,(q, r)$ and $b = h\,(s, t)$ permit the elimination of a and b from the descriptive function.

Obviously, a more adequate analysis than that of Meehl and Sellars is called for: nonetheless their paper, and that of Pepper, succeed in showing where the real difficulties lie and give various clues to the direction of that sufficient treatment.

Let us return now to the first part of Pepper's argument and the criticism of it offered by Meehl and Sellars. Pepper's argument begins with a supposition: the supposition that a function $f_1\,(q, r, s, t)$ adequately describes the interrelationships among the four variables q, r, s, t. The supposition is not questioned by Pepper's critics, although it would in fact seem that if it is granted then Pepper's conclusion must also be accepted: that $f_2\,(q, r, s, t)$ must be identical with f_1. Meehl and Sellars avoid the conclusion by claiming that the emergentist is speaking about different regions of 'space'-time. As we shall see, the emergentist's view is very different from what Meehl and Sellars describe it to be: but, leaving this aside, Meehl and Sellars in making this claim seem to miss the strength of Pepper's argument. Pepper, it would seem, is talking about a particular region of 'space'-time—including the four-space (q, r, s, t): he is talking, let us say, of the biochemistry of the brain or—to introduce an example we will find useful—of the biochemical variables and relations involved in the life of the amoeba, what we may call amoeba processes. His supposition is, then, that the amoeba processes are adequately described by such a function as $f_1\,(q, r, s, t)$ of a chemical order, a function perhaps of the type that theoreticians like N. Rashevsky have long been seeking.[18] We will gradually expose definite methodological reasons for doubting the possibility of finding such a function, but if its existence is granted then one cannot avoid Pepper's conclusion, that f_2 must be identical with f_1 by considering f_1 as applying to 'physico-chemical processes which are not occurring in protoplasm' and f_2 as applying to protoplasm.

Moreover, that consideration only postpones Pepper's conclusion by one step, as is evident from Meehl and Sellars' consequent puzzle: 'while the notion of different regions in the fourspace q, r, s, t exhibit different functional

---

[17] Ibid., 252.

[18] N. Rashevsky, *Mathematical Biophysics*, 2 vols., New York, Dover, 1960.

relationships is mathematically unexceptionable, is it emergence?'[19] No emergent variables have been introduced, nor, reapplying Pepper's argument here, can they be introduced. Meehl and Sellars succeed in introducing them in the odd manner already indicated, a manner of introduction which does not correspond to scientific practice in the question of emergent qualities, no more than does the introduction of different regions of four-space as representative of the emergentist's approach. The authors in fact would seem to be driven to these positions by the acceptance of Pepper's basic assumption, and so it would be as well to examine thoroughly that assumption before exploring how precisely emergent qualities or variables are introduced, what precisely the emergentists are claiming.

Pepper's assumption is indeed the heart of the matter, and its criticism here will rest fundamentally on our previous discussion of the nature of randomness and its opposition to system. There the question was discussed on the level of pure mathematics, and we saw how the random series could not be systematized even by functions of the type considered by A. Church[20] and that the only kind of systematization possible was that provided by a statistical understanding. Here the discussion moves on the more difficult level of empirical science, in particular on the level of cell biophysics and biochemistry. Our negative conclusion will be similar to the conclusion on the mathematical level: that the complex process— we take as example the amoeba process—resists in principle systematization on the level of physics and chemistry, that from the physico-chemical point of view it is really random, it is a coincidental aggregate of physico-chemical events as Lonergan puts it.[21] We may note here that after our discussion we will be in a better position to make more precise the meaning of systematization and randomness in general: our earlier mathematical considerations, however, leave them sufficiently defined for the moment. With our negative conclusion goes a positive conclusion which has no parallel on the mathematical level: the conclusion that the randomness on the biophysical and biochemical level is a necessary and, in a certain sense, a sufficient condition for systematization by means of biological laws.

First of all let us consider in detail the grounds for the negative conclusion. We take first the growing body of work which comes under the name of biophysics, and as spokesman of this approach we may take N. Rashevsky . As a concrete example let us consider his work, *Mathematical Biophysics,* especially that large portion of the first volume where he discussed the 'Mathematical biophysics of vegetative cells and cellular aggregates,'[22] for such a discussion is obviously near to our own example, the physico-chemical understanding of the amoeba.

---

[19] M. and S., 246.

[20] Cf. chapter two, footnote 17.

[21] Cf. *Insight*, 255–6; 262–4; CWL 3, 280–81; 287–89.

[22] Rashevsky, *Mathematical Biophysics*, Vol. I, 7–372.

Not that we ascribe any special authority to Rashevsky 's book. We might well have discussed work more closely related to recent developments in quantum biophysics or open-system thermodynamics.[23] As Rashevsky admits in the preface to the 1959 revised edition, the book might well have been totally rewritten rather than enlarged by the inclusion of later work. Still, the book as it stands is a convenient survey of this type of investigation, and the principles governing other such work are clearly operative in Rashevsky 's book. Moreover, if the thesis stating the irreducibility of biology to physics and chemistry is not to be rejected by an appeal to future physics and chemistry,[24] that thesis must be manifestly based, not merely on present physics, but on the method or principles governing these sciences.

Rashevsky 's book gives two main approaches to the problem of cell division. There is what may be called the continuum approach, the method of which is most evident in chapter one, where a diffusion equation is derived by the procedure which Lindsay and Margenau call 'the method of elementary abstraction.'[25] It is a method which parallels the standard approach to hydrodynamics when no account is taken of the actual discontinuities, molecular structure, etc., of the liquid: one which can be recognized as in principle failing to 'approach reality asymptotically, by gradual approximation.'[26] Thus the discussion in the earlier chapters of Rashevsky 's book of the problem of a stationary diffusion field in the case of a spherical cell in a liquid medium, with various simple types of substance consumption, and production, calls for obvious abstractions from the concrete object whose explanation is sought. Moreover, even on this level of abstraction, results are practically unobtainable without the constant introduction of averaging processes.[27] Nonetheless, the diffusion drag force theory, developed in chapter eight, is remarkable in giving some account of cell division, even giving the correct order of magnitude for the average size of living cells and correctly representing the dependence of the size on the rate of metabolism. However, the theory does not give even a generic account of mitosis, much less of the multitude of micro-reactions of the living cell. As Rashevsky points out,[28] it is clear that an extension of mathematical biophysics into the domain of molecular and atomic physics is called for. Not much of Rashevsky 's

---

[23] For example, in the line of H. Kaeser's 'Some Physico-chemical aspects of Biological Organisation,' 191–249 of *The Strategy of the Genes* by C. H. Waddington, London, Allen & Unwin, 1957.

[24] Cf. for example, Nagel, *The Structure of Science*, 363; Paul Oppenheim and Hilary Putnam, 'Unity of Science as a working hypothesis,' *Minnesota Studies*, II, Minneapolis, 1962, esp. 16–23.

[25] Lindsay and Margenau, *Foundations of Physics*, 29–47.

[26] Rashevsky, Vol. I, 1.

[27] Op. cit., Vol. I, 12–13; 16–17; 67, 89, 129, etc.

[28] Op. cit., Vol. I, 356.

book is devoted to this topic,[29] but sufficient to reveal its limitations. Inevitably, approximations and averaging are called for, and the results are general: for example, there is the series of inequalities given by equations (1) to (8) of chapter nineteen governing the general conclusions whether or not there will be cell division, whether or not there will be mitosis, etc. No doubt one may point out that this procedure of approximation, etc., is merely a practical one and that in principle the total process for any cell (or amoeba) might be calculated out in complete detail. This point in fact can be conceded without damage to our argument: for our argument at no stage rests on the premise of there being indeterminacy at any level. So, for example, there is no implication that there is something indeterminate about the motions, etc., of the molecules, etc., within a cell—no more than that the motions of the atoms in an ideal gas are indeterminate. They are precisely determinate: each molecule moves in a determinate fashion according to the laws of physics and chemistry: the sequence of equations of its motion could be written down with the relevant boundary values and collision conditions. Still, to be able to write down equations in anything like manageable form one has to avail of a certain amount of abstraction. Thus, one could not take into account the attractive forces of all the elements of the system: to do so would be to draw the entire aggregate of elements into each equation of motion, giving a system of equations that could scarcely be explicitly written down, much less solved in detail. Again, every collision, every concrete intersection of causal chains,[30] calls for a new set of equations, a determination of conditions, etc. Hence all along the line the same process must be used: if the system is to be conveniently handled, one must use ways of avoiding the inclusion of aggregates: of aggregates of masses, forces, etc.; of aggregates of sets of equations: of aggregates of conditions. One could, of course, write down in some symbolic way the total sequence of sets of equations and conditions, but this would only serve to make manifest the fact that we are dealing with non-systematic aggregates of equations, each involving non-systematic aggregates of conditions, etc. We may note too that further complications would be revealed by considering the presuppositions of the equations, the abstract laws of geometry and of motion. Rashevsky remarks in his introduction that 'only a superman could at once grasp mathematically all the complexity of a real thing. We ordinary mortals must be more modest and approach reality asymptotically, by gradual approximation.'[31] Now here we are dealing neither with a superman's grasp nor with the knowledge of a Laplace's demon, but with ordinary human intelligence. We are interested in seeing why it is that 'following the fundamental method of physico-mathematical sciences, we do not attempt a mathematical

---

[29] Op. cit., Vol. I, mainly chs. 19, 28.

[30] Cf. M. Bunge, *Causality*, section 5.

[31] Rashevsky, Vol. I, 1.

description of a concrete cell, in all its complexity.'[32] But why that attempt is futile can be appreciated only in so far as the subject, the philosopher, experiences the futility of the effort personally and moves to some introspective understanding of the basis of that futility—a basis surprisingly related to Aristotelian notions. To make that basis and experience indicatively available here would lengthen our treatment considerably, but it cannot be dispensed with: one must follow through, introspectively, something like Rashevsky's efforts if one is to have the elements of a precise appreciation of the differences and the relations between the unsystematic understanding given by the lower sciences and the systematic understanding of biology. We anticipate in this remark the fuller discussion of chapter twelve.

Our case rests on the fact that a mathematical or physical understanding of the concrete cell would not be systematic, would not be expressible in a uniform set of equations from which the whole process might be deduced, given some basic sets of conditions. Undoubtedly one may visualize a way of unifying such equations by using functions of the type proposed by A. Church to deal with random sequences, so that if the equations for various stages were given, with definite values for all coefficients and conditions, one could produce a set of neat formulae from which one could generate these values. But for many reasons such a unification offers little advantage. As already pointed out, such functions are an *a posteriori* ordering of values. Also in the present case the order in which the equations, for any given time, and their components are laid out is irrelevant to the problem. Again the values involved are values of particular forces, particular coupling constants, etc., so that even if a generating formula for a matrix of numerical values were available, one would require a specification of the matrix loci—a specification equivalent to providing correspondence rules—which specification would be a reintroduction of aggregates which fluctuated randomly. There is too the basic oddity that the number of equations and the dimensions of the matrix of values involved are dependent on the precise number of elements— say molecules—of the system: a change in the latter number, even by one, calling for a recasting of the entire set of equations and matrix. This last point is most evidently related to the fact that 'we do not attempt a mathematical description of a concrete cell, in all its complexity.' The attempt is not made, not because we are not supermen, but because somehow such an attempt would not be essentially relevant to the understanding of the cell or the amoeba. As H. Kaeser remarks:

> It is well known that organisms consist of molecules. It should, therefore, be possible to account for biological behaviour in terms of molecular behaviour. Yet it is evident that the complete enumeration,

---

[32] Ibid.

even if it were possible, of all the molecules within an organism would not account for any but its most trivial aspects.[33]

Here, however, we have come to a position of being able to pin down the precise reason for this failure. Let us note, then, how far removed our proposed sequence of sets of equations with conditions would be from S. Pepper's simple $f_1(q, r, s, t)$ which was to account, say, for the amoeba on the non-biological level. The counterpart of Pepper's f_1 in our account would be the sequence of sets of integrals of the equations, with boundary conditions specified and included— conceding for the sake of simplicity of argument that the integrals of this many-bodied problem could be made available. It is to be noted in particular that a sequence of sets of equations and conditions are required—e.g., the set changing after each collision or reaction—and that this sequence is random, non-systematic. The central point then is the random or coincidental nature of the proposed account: the non-biological account at best involves a random, or non-systematically related, set of functions of a randomly-varying set of variables. Taking Pepper's f_1 to symbolize this complex but determinate account, we may note that he is right in claiming it to be unique. Both Pepper and his critics centre their discussion on the question of uniqueness: but it is not uniqueness that is significant or relevant to the problem of emergent qualities, it is complexity and the precise nature of the complexity. From the point of view of mathematical biophysics the processes involved within the amoeba form a coincidental aggregate which can be understood concretely only through a coincidental aggregate of equations and conditions.

Nor is this the only difficulty. For, the result of such a thorough investigation of any one cell or amoeba would be applicable only to that cell or amoeba. What Rashevsky remarks in regard to the continuum approach holds also for the molecular approach:

> Even granting that we could solve exactly the differential equations of diffusion for a large number or even for all cases, this would still leave us with a rather serious handicap. The distribution of the diffusion flows in every individual case depends, among other things, on the exact shape of the cell. A slight variation of the latter will modify the analytic expressions describing the distribution of concentrations and flows. But, since there are no two cells perfectly alike, the exact solution of the problem for a given case would contain a tremendous amount of detail which is biologically insignificant because it applies *only to the given case.*[34]

---

[33] *Biological Organisation at the cellular and subcellular level*, ed. R. Harris, New York, Academic Press, 1963; H. Kaeser, 'The Kinetic structure of Organisms,' 25.

[34] Rashevsky, Vol. I, 15; italics his.

Rashevsky suggests two ways of overcoming this difficulty. Either one solves the problem exactly in a single case hoping that other cases will fit approximately this solution, or one concentrates on 'such gross features as are common to all cells of a given type, in spite of a difference of detail.'[35] The difficulties of the first course have been touched on above: exact solutions are forthcoming only in so far as one simplifies the equations, thus departing from the concrete problem; but even if one could reach a non-simplified solution of a particular problem the solution in its exactness will be a non-systematic aggregate of physical equations and conditions, referable only to the particular case. We are led then to try the second method, for

> the individual variations from cell to cell are so large that it would be futile to compare exact numerical values. But the regularities that are found in the behaviour of cells, in spite of these individual variations, suggest a search for relations of a general form which may remain invariant from cell to cell despite variations of numerical values.[36]

One thinks immediately in this context of equations invariant in structure,[37] or of topological relations[38] or of open-system thermodynamics.[39] But all these approaches involve obvious abstraction from the concrete process being studied.[40]

It seems, then, that on the level of physics either one abandons hope of generalization and attempts the endless task of a complete solution of each particular problem, each 'amoeba process' involving coincidental aggregates of equations and conditions, or one reaches generalization only by abstraction from relevant aspects of the process.

Still, besides the biophysical level there is the biochemical and recent advances on this level might lead one to suspect that one could overcome the difficulties of randomness and complexity by utilizing these advances. Here we will restrict ourselves to general considerations of the biochemical explanation of 'amoeba processes.' There are of course many specialized discussions of particular aspects of amoeba processes, such as those on the problem of amoeboid movement, but these would only serve to strengthen our thesis: that, from the biochemical point of view, the processes of metabolism and cell division are random, coincidental processes, and the biochemical account of such processes cannot but be random or non-systematic.

---

[35] Ibid., 16.

[36] Ibid., 37–38.

[37] Ibid., 37.

[38] Rashevsky, Vol. II, 245–423, passim, and more recent references given there.

[39] Cf. footnote 23 above: also K. G. Denbigh, *The Thermodynamics of the Steady State*, London, Methuen, 1951.

[40] How this is overcome on the biological level will gradually emerge.

The most obvious difference between biophysical and biochemical discussion of amoeba processes is the consistent utilization in the latter of new units on the molecular level with their properties of structure, affinities, bond energies, etc. This introduction relieves one—to some extent only, for physical properties are still relevant to biology—of handling the details, states, etc., of sub-molecular elements. So, for example, the ordinary account of DNA does not mention electron configurations: not, to repeat a point already made, that the electron configuration is indeterminate, or not playing a determinate role; but the determination and the determinate role in any one cell is non-systematic, and also it differs from cell to cell in random fashion, in a fashion which for chemistry is empirically residual. On the level of physics, indeed, the aggregates named DNA differ one from another: a physical account of one such aggregate would be strictly inapplicable to any other and would itself be a non-systematic aggregate of equations, etc. Yet in the reactions of these aggregates with like aggregates, e.g., RNA aggregates, regularity or system is manifest and this appearance of system warrants the systematization given by chemistry, where the basic units of systematization are atomic and molecular, and their basic correlations are given implicitly by the ordering of the periodic table and explicitly in reaction equations. We are not, however, concerned here directly with the distinction between physics and chemistry, nor with the fact that, as Cassirer remarked, the notion of chemical element is indisputable.[41] We wish rather to draw attention to the relation of randomness to the possibility of such a distinction. In doing so we succeed in specifying the meaning of randomness at various levels much as we specified that meaning earlier on the mathematical level. Just as the random series on the level of mathematics escapes systematic treatment or—in suitable cases defined by Von Mises' axioms—falls under statistical generalization, so on the level of physics explanation may involve random aggregates of equations and conditions which cannot be systematized on that level. They can, under certain conditions, be in a certain sense systematized through statistics—one thinks immediately of thermodynamics—and as we have just noted, there is also the possibility that the random aggregates, to which the physical equations refer, be systematized by the use of chemical notions and equations. Thus randomness appears as the condition both of the objectivity of statistical science and of the possibility of the systematization at another level, a systematization intimately connected with the emergent laws and qualities mentioned earlier. The standard reductionist reaction to a possible distinction between physics and chemistry is to point out the possibility of deriving through quantum mechanics a complete account of the states of any given molecule. But—even laying aside the fact that

---

[41] E. Cassirer, *Substance and Function*, New York, Dover, 1953, ch. IV, section 8; F. A. Paneth, 'The Epistemological Status of the Chemical Concept of Element,' *British Journal for the Philosophy of Science*, (13), 1962–3 (original 1931).

quantum theory deals not with processes but with values and distributions—the real crux of the matter is, whether the account on the physical level is systematic. Thus, going a step further, if the complete physical description of the amoeba and its processes were systematic, yielding a neat unified set of equations such as for the ideal solar system, then any other systematization would certainly be superfluous. But it is not enough to point to the determinacy of any one level to exclude a systematization on a higher, more inclusive, level.

Even at this stage the importance of the notion of randomness for a methodological understanding of evolution theory should be evident. For, evolution theory involves considerations both of statistics and of emergent laws: without objective randomness, in the sense we have been describing, the statistical generalizations involved in evolution theory would be a mere temporary cloak of ignorance, and emergence in general would be, to use Pepper's term, an epiphenomenon. Bertalanffy to some extent adverts to this importance when he discusses science as a hierarchy of statistics, but the situation is not as simple as he makes it out to be. Firstly he exaggerates the role of statistics: 'all laws of nature are of a statistical nature. They are statements about the average behaviour of collectives. Science as a whole appears as a hierarchy of statistics.'[42] Secondly, the hierarchy is not one of statistics but of randomness. The non-systematic of one level is systematized on the next level without contradiction. Various authors[43] have stressed the existence of integrative levels, but have failed to specify the possibility of such integrative levels, the possibility challenged by Pepper's paper. To specify it requires advertence to the distinction between the complete concrete non-systematic account on one level and the inclusive systematization or integration of the higher level. With this distinction understood, Feibleman's view that 'an organisation at any level is a distortion of the level below'[44] appears groundless. As is commonly held, but without theoretical justification, the lower level laws are not violated, nor are its processes distorted. Rather, the non-systematic aggregate of determinate processes of one level are systematized at the other level and display a certain degree of empirical residuality. From another point of view, as Feibleman puts it,[45] the mechanism of the higher level lies at the level below, and it is this fact which makes the reductionist view plausible. Feibleman also remarks on the various relations of interdependence between the levels, but we will postpone a consideration of these till we have dealt with the level of biological relations.

Let us consider now more directly the biochemical account of cell structure and function. At present attention in the field is centred around the nucleic acids

---

[42] Bertalanffy, *Problems of Life*, London, Watts & Co., 1952, 172.

[43] Cf. J. K. Feibleman, 'Theory of Integrative Levels,' *British Journal for the Philosophy of Science* (4), (1953–4), 59–66.

[44] Ibid., 63.

[45] Ibid., 61.

in their relation to metabolism and to meiosis, etc. There is even popular interest in the question of the genetic code, and in a standard text on the subject we find the general attitude on a more academic level expressed in chapter headings: 'The nuclei acids, carriers of biological information,' 'The ribosomes and the utilization of information,' etc.[46] As this general attitude and the concomitant linguistic usage exemplified here are a stumbling block to the appreciation of what precisely is going on in cell biochemistry we had best make some effort to get behind it. One might do this conveniently by considering the peculiar problem of inventing something like an amoeba, i.e. a unit which ingests, grows, multiplies indefinitely, etc. Granted that the inventor had some notion of what he was trying to invent, there would be the obvious need of finding parts suitable for the preservation and multiplication of pattern—just as the inventor of the watch had to seek out a source of regulated energy. To use an old Thomist tag, the required parts are parts disposed to the form, the pattern of structure and motion. While there is certainly a tendency which inclines one to say that the parts specify or fix the pattern, in fact it is more true to say that the pattern specifies or organizes the parts, somewhat as the form of a house gives organization to the bricks. It is the prior inclination which leads one to say 'the nucleus is the repository of hereditary information,' 'DNA is now recognized as being the chemical structure that stores the cell's hereditary information.'[47] Undoubtedly DNA plays an essential structural role and an essential role in continuity: but the role is more like the role of resources and terrain in determining the pattern of a battle than like the role of a general's plan. One may think here of the role of spring and escapement in the watch and their different contributions to the phenotypic watch. On the biological level, the phenotype would seem normally to involve description of biological function, while the genotype is explanatory of the biochemical form of the lower aggregate in the nucleus with certain relations to the described phenotype.[48] Again, different types of parts give different watches, so that watches may be distinguished by the structure of their parts, but the parts contain the plan of the watch only in a particular potential sense. The biological parallel is obvious, and the purpose of the parallel and the discussion is to reveal the danger of the use of the language of cybernetics in biochemistry. One may speak of the correlation of twenty amino acids with sequences of bases as a problem of decoding, but in doing so one must guard against the tendency to consider the code as involving a microplan of the organism.

---

[46] Loewy and Siekevitz, *Cell Structure and Function*, New York, Holt Rinehart and Winston, 1963: the titles of chapters 7, 13.

[47] Ibid., 56.

[48] 'The base triads are not the adult characters considered from another point of view, in particular it would be wildly absurd to call them the adult characters redescribed!' R. Harré, *Matter and Method*, 23.

This is not to say that the chromosomes and genes have no part in determining the type and function of the organism: to say this would be worse than to claim that pure chemical elements given as food in the correct proportions would be a suitable diet. Chromosomes, etc., are in a basic sense a material component of the organism, but they are not 'raw' inert material. Waddington comes closer to specifying their role when he remarks

> we can say that the hereditary materials with which an organism starts off define for it a branching set of creodes. A path of development, or creode, exhibits a balance between inflexibility (tendency to reach the normal end result in spite of abnormal conditions) and flexibility (tendency to be modified in response to circumstances) ... The creode is an expression of the potentialities derived from the whole set of genes.[49]

From another point of view one may say that the hereditary material limits the range of possibilities in a particular organism: but all these words 'limitation,' 'determination' involve an ambiguity which gives rise to

> a rather paradoxical situation about organization. In a way it is a limitation on the potentialities of the system. You have a lot of units or elementary parts, and if they fit in a certain defined pattern, that eliminates the possibility of their being joined up in other ways. On the other hand, it is only by this fitting together, even though it is in a limited way, that the full range of properties of the elements ever emerges and becomes manifest. In fact organization, although a limitation in one way, often means that the units are exhibiting properties which they cannot show in isolation.[50]

To put the matter in a broader context and in a novel way one may say that it is only through the operation of the total environment the genes operate, where 'operate' has the ambiguity of 'determination' and of 'being determined.' The spring within a watch plays an essential but dependent part, but, unless the watch is waterproof, it will fail under water: or, to take an example more in line with the entropy conditions in the organism, a photo-cell dependent mechanism will be useless if kept in the dark.[51] In this sense too, 'any one genotype may give rise to somewhat different phenotypes, corresponding to the different environments in which development occurs.'[52]

---

[49] C. H. Waddington, *The Nature of Life*, London, Allen & Unwin, 1963, 63; a more elaborate account is given in his *Strategy of the Genes*, London, Allen & Unwin, 1957.

[50] *Biological Organisation*, Oxford, 1959, ed. C. H. Waddington; Introduction, 2.

[51] Cf. Albert Ducrocq, *La Logique de la Vie*, Paris, 1956, esp. ch. V, 'Cybernetique et Biocybernetique.'

[52] Waddington, *The Nature of Life*, 29.

We have stressed the ambiguity of the word 'determination' and the distinctions necessary in talking of heredity determinants, codes, etc., in order to lead gradually to what seems to be the only genuine solution to the problem of duality which recurs at all levels. Bertalanffy remarks on this dualism:

> Biological laws are not a mere application of physico-chemical laws, but we have here a realm of specific laws. This does not mean a dualism in the sense that vitalistic forces enter into play in the living. But it appears that the level of biological laws is a higher one, as compared to that of the laws of physics.[53]

Bertalanffy denies dualism, but his efforts to explain its appearance or to exclude it, e.g., by generalized constructs referring to isomorphisms between the levels[54] have from various points of view been found wanting.[55] The relevant solution to the problem has, in fact, been adverted to by more than one author, e.g., by Waddington:

> When it turns out that certain arrangements of the atoms of carbon, nitrogen, oxygen, etc., exhibit properties which we recognize by the name of enzymes ... it is completely out of the picture to suggest that we have to add something of a nonmechanical kind to an already fully comprehended material atom ... The secret of their performance in this way is architecture, or to use the Aristotelian term, form.[56]

But it is not sufficient merely to recall an outworn terminology of matter and form: the Aristotelian couplet must be properly and contemporarily conceived, and this can be done only in so far as the standard paradigm cases for its understanding—especially as presented by scholastic philosophers—are replaced by the fundamental paradigm case and the origin of the couplet. But we had best complete our discussion of the way in which biological relations are supervenient on the physico-chemistry of the cell, etc., before tackling this general question of genuine supervenience without duality. We will, however, do this in such a way as to prepare the ground for an appreciation of the modernized version of the Aristotelian couplet, matter-form, to follow.

All along our attention has been centred on the opposition of system to absence of system: the opposition between the systematic and the random or non-systematic. Neither the systematic process nor the non-systematic process leave any scope for indeterminism: the solar system is perfectly determinate, but so also is the movement of traffic in Oxford. On the other hand, the solar system

---

[53] Bertalanffy, *The Problem of Life*, London, Watts, 1952, 157.

[54] Bertalanffy, 'The Mind-Body Problem: A New View,' *Psychosomatic Medicine,* (26), 1964, 29–45.

[55] John Lachs 'Von Bertalanffy's New View,' *Dialogue*, 4 (1965–6), 365–370.

[56] Waddington, *The Nature of Life*, 21.

is a suitable candidate for Laplace's type of determinism: but no knowledge of morning conditions in and about Oxford would enable one to predict the state of the afternoon traffic. Not that in principle prediction is impossible in the case of non-systematic processes. One might well patiently work from the first significant set of equations and conditions to their first failure or discontinuity, elaborate a further significant aggregate of equations, etc. But the deduction would be no more systematic than the situation being studied. It would be piecemeal, non-unified, involving—as we saw so well in the case of the biophysical explanation of cell processes—a sequence of coincidental aggregates of equations and conditions.

Again, our attention has been focused more on our understanding of various situations than on the situations, and so we have distinguished systematic and non-systematic processes by the type of understanding required to master each. Both processes require concrete combinations of abstract laws and initial conditions. In general one can know already, from the defined initial or basic situation, with which type of process one is dealing:

> If many different and unrelated insights are needed to understand the basic situation, the premises for a deduction from that situation cannot be a single, unified combination of selected laws. And since a coincidental aggregate of premises will yield a coincidental aggregate of conclusions, it follows that every deducible situation, provided it is a total situation, also will be a coincidental aggregate. Further, it follows that, when a non-systematic process happens to give rise to a systematic process (as in recent theories on the origin of planetary systems), then the total situation must divide into two parts of which one happens to fulfil the conditions of systematic process and the other fulfils the requirement of other things being equal.[57]

We may recall here Pepper's assumption of the existence of a function f_1 (q, r, s, t) to describe a given situation in biochemistry. In the light of the foregoing discussion it should be evident by now that such an assumption could only hold for a systematic process.

There is a counterpart on the level of imagination to this existence or nonexistence of a unified explanation, evident if one compares one's efforts to picture the solar system to the effort to picture the processes of even an ideal gas. Somehow only the systematic process is 'imaginatively controllable.' There is a parallel situation too on the level of model construction. Models for the planetary system are well known, but making models for the ideal gas would seem pointless, and to make the latter type of model the key rule would be to make sure that

---

[57] B. Lonergan, *Insight*, 50; CWL 3, 74.

irregularity, absence of rule, was allowed to enter into the composition of the initial situation.

Now in dealing here with the biochemical level we will not retrace the steps taken, when we discussed the biophysical level, to make manifest the various random aggregates involved in the lower level account of the cell. We will stress rather the aspects mentioned in the last paragraph, particularly in order to bring out the relation of imaginative representation of biochemical process to the understanding of biological relations. In a lengthier discussion one might consider in detail, for example, the standard text book presentation of cell structure and function.[58] Such a presentation normally begins with a basic general description[59] and a schematic diagram of the cell,[60] then it passes to a detailed discussion of various particular processes: the synthesis of proteins, transport problems at membranes, etc. This fragmentation of attention is of course related to the non-systematic nature of the chemistry of the total cell process. Towards the end of the book one has reached a better understanding of the cell on the biochemical level and concluding diagrams,[61] symbolizing not merely the static cell but its dynamics, enable one to grasp more clearly the biological relations involved. In actual fact, indeed, the biological relations are implicitly acknowledged all along, and what the text actually attempts is 'to trace the roots of biological process into the known realms of physical and chemical phenomena,'[62] thereby better locating and more precisely defining such relations. But so close is the interrelation of biochemistry, so close is the interplay of diagram and understanding, that one can easily be tricked into regarding the biological relations as either absent or superfluous, 'epiphenomenal,' and if one turns the trick into a philosophical position one is a reductionist.

> Suppose I am trying to explain why such an organism can reproduce. Am I to allow the law "molecules combined in just such a manner can reproduce" as a law of Physics? Certainly not! The phrase "combined in just such a manner" surely describes a biological property, and hence we achieve reduction to Physics only at the cost of making Biology a part of Physics by a trick.[63]

---

[58] We have already referred to such a book in footnote 46; we will continue to use it as an illustration.

[59] Op. cit., footnote 46, ch. 3.

[60] Ibid., 18.

[61] E.g., ibid., 158–220, with the diagrams there.

[62] Paul Weiss, *Principles of Development*, New York, Holt, Rinehart, and Winston, 1939, 108.

[63] J. G. Kemeny, *A Philosopher Looks at Science*, Princeton, NJ, Van Nostrand, 1959, 214: for a more accurate statement cf. the quotation from Woodger in H. Hermann,

On the other hand if one is critical, one comes to acknowledge that 'the chemistry of the cell can yield an image of catalytic process in which insight can grasp biological laws.'[64]

The biological laws are to be distinguished from other laws which may equally well be grasped in such images: for example, laws of topological equivalence. Unlike the topological laws, they are appreciated precisely as systematizations of the chemical processes.[65] We cannot altogether agree with the view of F. A. Hayek but his expression comes close to ours:

> The biological relations are explanatory of the concrete chemical processes and, inversely, the understanding of the chemical processes is explanatory of the mechanism of the biological process. Biological processes, constructed from biological laws, make systematic what would otherwise be unexplained regularities in the non-systematic chemical processes.'[66]

This inclusion of the lower level is flexible, and it is this which leads Bertalanffy to remark that

> the processes in the living are so complicated that with laws concerning organic systems as a whole we cannot take into account the individual physico-chemical reactions, but use units and parameters of a biological order ... we have to use constants which are expressions for the lump result of innumerable physico-chemical processes[67]

and that 'nearly homogeneous material—nerve cells, muscle cells, bone cells—builds a typical formation, the outer shape of which is characteristic and specific, while the inner arrangement of the cells is largely accidental.'[68]

Thus, returning to our simple example, in the diagrammatic representation of the coincidental aggregate of chemical amoeba processes[69] one may grasp the biological processes—with their implicit relations—of reproduction, ingestion, digestion, etc. Without the biological correlations one is left with aggregates of

---

'Biological Field Phenomena: Facts and Concepts,' *Form and Strategy in Science*, edited by J. Gregg and F. Harris, Dordrecht, D. Reidel Publishing, 1964, 351.

[64] Lonergan, *Insight*, 256; CWL 3, 282.

[65] This thesis will emerge from the remainder of the chapter. A fuller treatment of the relation between chemistry and botany in particular is given in the paper cited above in footnote 2, lxiv.

[66] F. A. Hayek, 'The Theory of Complex Phenomena' in *The Critical Approach to Science and Philosophy*, Edited by M. Bunge, New York, Free Press, 1964, 336.

[67] Bertalanffy, *The Problem of Life*, 154; cf. also 172.

[68] Ibid., 63.

[69] Or of course, more directly, but less exactly, in a microscope.

chemical equations or abstract topological relations. Nor has the biologist much difficulty in verifying the occurrence of biological system in particular cases: the odd amoebid properties of chloroform or of alcohol-injected clove oil do not deceive him.[70]

But a more precise analysis of the definition of these biological relations is required, and some account of their relation to the levels of physics and chemistry. The first thing to note about the biological level is its apparent definitional unconnectedness with the lower levels. I use the word 'apparent' because certainly the meaning of the biological relations does involve lower level 'mechanisms' and it is for this reason that one seeks 'to trace the roots of biological process into the known realms of physical and chemical phenomena.'[71] But in fact biological relations are defined in terms of each other. One may recall here the procedure of implicit definition, most evident in mathematics: for example, point and line as defined in various axiom systems. It is this interdependence of definition which underlies the difficulty in defining the kidney, alluded to by Beckner:

> It is apparent that what one ordinarily thinks of as a morphological description, i.e. in terms of colours and shapes, cannot be a definitive description of the kidney. A much closer approach to such a definition may be achieved if general bodily position of the organ and the paths of its ducts are taken into account. Here one would have to beware of eventual circular definition, for instance, defining "kidney" in terms of "abdomen" and defining "abdomen" in terms of the organs it contains. In practice, a functional definition is the obvious way out. Any structure is said to be a kidney only if it contributes to a function F^1, in this case excretion.[72]

But how then is one to define excretion? Beckner in fact cannot escape circularity: like the parts of the body, the definitions of the various functions interlock. The point is more explicitly made in Meehl's paper 'on the circularity of the Law of Effect,'[73] where he draws an interesting parallel between the circularity of the Law of Effect and Mach's treatment of force and mass. On the level of psychology in general there is the definitional interdependence of conscious stimulus and response. 'In any set of psychodynamic principles a basic vocabulary is necessary. Such a vocabulary will now be operationally derived from two observables or "facts"—perception and response ... Perception and response would appear to

---

[70] Cf. Bucksbaum, *Animals Without Backbones*, Vol. I, Pelican, 1951, 30–31.
[71] Cf. footnote, 62.
[72] M. Beckner, *The Biological Way of Thought*, 116.
[73] Psychological Bulletin (47) 1950; reprinted in *Readings in the Philosophy of Science*.

be basic, unarguable characteristics of living organisms.'[74] This paper contains a reformulation of Freudian psychoanalytic principles through the use of more operational, observational terminology and it exposes an important point about which the reader may have already wondered: the fact that the stress all the time in our present discussion is on relations rather than on entities or terms. The answer to this is partly contained in Ellis' statement that 'the existing orthodox Freudian principles make use of many vague terms like "libido," "id," "energy cathexis," etc., which may be legitimate hypothetical constructs but are most difficult to pin down to earth and experimentally validate or disprove.'[75] Closer still to our point is Skinner's criticism:

> What has survived throughout the years is not aggression or guilt, later to be manifested in behaviour, but rather patterns of behaviour themselves. It is not enough to say that this is "all that is meant" by sibling rivalry or by its effects upon the mental apparatus. Such an expression obscures, rather than illuminates, the nature of the behavioural changes taking place in the childhood learning process.[76]

The point is, not that operationalism is correct, but that meaning on any level is relational. On the level of mathematics this was adverted to by Pasch almost a century ago when he pointed out that the deductions in axiomatic geometry should be independent of the meaning of the geometrical 'concepts,' that only the relations specified in the propositions and definitions employed should be taken into account.[77] The meaning of the 'concept' or term is in fact in the relation. It is, indeed, a matter of common experience that, when we attempt to define anything, unless the 'definition' is merely ostensive, our efforts to explain the thing will be directed to specifying its relations. For this reason I may make my own the words of C. L. Morgan:

> The discussion of relatedness ... requires the consideration of the terms in relation within any given field of relatedness, and of the relations of these terms. Relatedness, in my sense of the word, includes both; not the terms only; not the relations only; for they can never be divorced if my usage of the word "term" be provisionally accepted ... On this understanding what is supervenient at any emergent stage of

---

[74] Albert Ellis, 'An Operational Reformulation of Some of the Basic Principles of Psychoanalysis,' *Minnesota Studies*, Minneapolis, 1962, Vol. I, 137.

[75] Ibid., 136.

[76] B. F. Skinner, 'Critique of Psychoanalytic Concepts and Theories,' *Minnesota Studies*, Minneapolis, 1962, Vol I; 84; cf. also 86.

[77] M. Pasch, *Vorlesungen über neure Geometrie*, Leipzig, 1882, 98–99.

evolutionary progress is a new kind of relatedness—new terms in new relations.[78]

In as far as one prescinds from the defining relation, one is no longer thinking of the term as defined: one is indeed probably tending in the same direction as Freud in constituting unverifiable entities. As Lonergan remarks,

> on many occasions Freud represents the outlook of his time and tends to regard observable events as appearances and unobservable entities as reality. What precisely is the libido? Is it what is known either by observing psychic events or by correlating these observables or by verifying these correlations? Or is it a construct that stands to Freud's verified correlations in much the same manner as the sponge-vortex aether once stood to electromagnetic equations?[79]

Lonergan's views on relations are more clearly developed than Morgan's and more in harmony with scientific practice, as is his view on objectivity.[80]

Also it is to be noted that not only do the relations involved in the understanding of any organism form an interlocking definitional pattern but their fuller definition requires the correlations of comparative and evolutionary studies. Beckner lays stress on this comparative meaning which may be applicable over a wide range of taxa: for example, the process named 'escape reaction' occurs in practically the whole range of living things, and it involves in each case somewhat different sets of biological relations.[81] One may well compare the correlates of evolution theory with the correlations given on the chemical level by the periodic table. Thus indigestion, respiration, reproduction, etc., not only are relationally defined among themselves in respect to a given organism: they also find further meaning through comparative and evolutionary correlations.

The view proposed of the sciences as layers of interlocking relation systems clashes with the view implied in Hempel and Oppenheimer's claim that, to obtain a less trivial interpretation of the emergentists' assertion of irreducibility

> we have to include in the explanatory theory all those laws known at present which connect the physico-chemical with the biological "level," i.e. which contain, on the one hand, certain physical and chemical terms, including those required for the description of

---

[78] C. Lloyd Morgan, *Emergent Evolution*, 18; chapter 3 of the book deals in detail with relations.

[79] Lonergan, *Insight*, 204–5; CWL 3, 229.

[80] On objectivity compare Morgan, op. cit., 198ff. with Lonergan, op. cit., esp. chapter 13: and both these views with the normal scientific practice in verification.

[81] M. Beckner, *The Biological Way of Thought*, 116 ff.

molecule structures, and, on the other hand, certain concepts of biology.[82]

Without this inclusion, which Kemeny would call a trick, the reductionist task becomes, to say the least, difficult. Woodger puts this difficulty neatly: 'If from a chemical hypothesis C, containing no embryological set designation, an embryological statement E is to be derived as a necessary consequence, this will only be possible if C is conjoined with a definition stating that the embryological set designatum E are abbreviations of chemical set designations of C.'[83]

Our lengthy discussion of biophysical and biochemical processes as non-systematic on their own level but systematized by biological relations, and of the interlocking definitions of biological relations and processes, should enable the reader to appreciate that the definition alluded to could not be legitimately included. For instance, the required definition in the case of 'digestion' would run somewhat as follows: 'digestion is an enzyme-catalysed transformation of carbohydrates, etc., etc.,' where all this is spelled out in chemical terms. The oversight, or trick, involved in such a definition has two aspects. The first is the systematic omission of the understanding of digestion which relates it as a whole to other processes such as reproduction, respiration, etc.; the exclusion of the fact that 'we can, on the one hand, isolate single processes and define them in terms of physics and chemistry. On the other hand, we can state overall laws for the biological system as a whole, foregoing determination of the individual physico-chemical processes comprised.' [84] It is not, however, so much that this understanding is not admitted: it is admitted without difficulty once the biologist turns back from philosophy to his laboratory. It is perhaps rather that it is too obviously present to be systematically acknowledged, especially if a philosophically-biased objectivity is accepted. From this point of view one might compare the biological correlations or axioms to the axioms of order in Euclidean geometry: it is not an easy matter to identify precisely what one is understanding in a diagram, whether the diagram is one of a geometrical problem or of biochemical processes. The second element overlooked in the proposed definition is the opposition of systematic to non-systematic. Digestion is understood as a systematization of a non-systematic aggregate of chemical processes. It is not an 'abbreviation,' such as one may get by using statistical averages; it is not an abstract generality, such as one may get from topology. It remains, however, to determine somewhat more accurately and generally what precisely is the relation between the two sides of the pseudo-definition.

---

[82] Hempel and Oppenheim, 'The Logic of Explanation,' reprinted from Phil. Sc. (15) 1948, in *Readings in the Philosophy of Science*, 336.

[83] Quoted by H. Hermann, 'Biological Field Phenomena, facts and concepts,' *Form and Strategy in Science*, 351.

[84] Bertalanffy, *Problems of Life*, 172.

To specify the relation various authors have, in fact, had recourse to the old Aristotelian distinction between matter and form,[85] but without any basic consideration or justification of the distinction. Thus Alexander: 'To adopt the ancient distinction of form and matter, the kind of existent from which the new quality emerges is the "matter" which assumes a certain complexity of configuration and to this pattern or universal corresponds the new emergent quality.'[86] Lonergan puts somewhat the same point in neater form: 'a concrete plurality of lower entities may be the material cause from which a higher form is educed.'[87] Elsewhere, however, Lonergan pinpoints clearly the paradigm for the distinction: 'The general analogy is the proportion of wood to tables and bronze to statues; but the specifically Aristotelian analogy is that natural form is to natural matter as intelligible form is to sensible matter, that is, as the object of insight is to the object of sense.'[88] If form and matter are not understood through this basic analogy with insight into phantasm, then it is more than probable that they are being misconceived. To grasp the correct analogy one must appreciate, for example, the relation between the cluster of clues and the solution in a good detective novel, or the distinction between the presentation and its significance in some of N. Hanson's diagram puzzles,[89] or the connection between the oval shape of the ellipse as seen and the ellipse as understood and so defined. Or, to come back to the problem on hand, the relation between the image of cellular, or amoeboid, chemical processes and the systematic understanding which expresses itself in biological terms. From this vantage point one can better appreciate the flaws in the effort to 'define' digestion in terms of chemical processes.

This is not to say that these processes are in no way relevant to an understanding of the biological level: for the chemical processes are what Aquinas would call the proportionate matter of the biological laws.[90] A man cannot be defined without bones: still, a man is more than bones. It is this notion that underlies C. L. Morgan's discussion of the involvement of one level in the next level, mounting up to the involvement of sense in understanding where we located the paradigm.[91] This involvement of one level with the next, the proper disposition of the lower level as matter, should not be conceived naively. One

---

[85] For example, Waddington, *The Nature of Life*, Cp. 2; Bertalanffy, *Problems of Life*, 150; Morgan, *Emergent Evolution*, 11ff; S. Alexander (cf. footnote 86).

[86] S. Alexander, *Space, Time and Deity*, London, Macmillan, 1920, Vol. I, 47.

[87] *Theological Studies* (IV), 1943, 'Finality, Love, Marriage,' 480. Republished in *Collection*, London and New York, Herder and Herder, 1967, 20; CWL 4, 20.

[88] *Verbum: Word and Idea in Aquinas*, 147; cf. also 43; CWL 2, 158; also 56.

[89] N. Hanson, *Patterns of Discovery*, Cambridge University Press, 1958, ch. I.

[90] 'The Concept of *Verbum* in the Writings of St. Thomas,' *Theological Studies*, X (1949), 7.

[91] C. L. Morgan, *Emergent Evolution*, 15–18.

might think of it in terms of the use of drugs for psychological disorders, or of the dependence on chemical change in the different states of muscle:

> There the addition of a "random substance," called adenosine triphosphate, to the highly organized nonrandom system of muscle fibre is responsible for the change in the state of the muscle fibre from relaxed to the contracted state. This shows a possibility of how a random substance can very readily and specifically influence the state of a highly organized system.[92]

It is obvious from illustrations such as solving a problem, finding a definition of an ellipse, that there is no logical transition between the various levels, where logic is understood in its normal sense. An interesting parallel is suggested here: just as there is no logic of discovery in this sense of logic, so there is no logic which could carry us from biochemistry to biology. To this point we return in the concluding chapter. But we may also note that an absence of linkage was not admitted: that linkage and mode of transition was specified by its relation to insight into phantasm as paradigm. That the same relation of insight to phantasm is central to the broader logic or discovery should not be surprising: 'hoc quilibet in seipso experiri potest, quod quando aliquis conatur aliquid intelligere, format sibi aliqua phantasmata per modum exemplorum, in quibus quasi inspiciat quod intelligere studet. Et inde est etiam quando aliquem volumus facere aliquid intelligere, proponimus ei exampla ex quibus sibi phantasmata formare possit ad intelligendum.'[93]

The closeness of the linkage is further manifested by the need to specify accurately the proportionate matter, here the biochemical processes, if one is to reach a clearly explanatory definition of the biological laws.[94] If one does not venture into biochemistry and biophysics one remains on the descriptive level in biology. A similar situation recurs on all other levels, for example, one cannot remain satisfied, in chemistry, with the view: 'Chemistry is a science in which interest is directed towards the secondary qualities of substances.'[95] The secondary qualities normally coincide with descriptive relations, smell, taste, colour, etc: to break through description here requires investigations on the level

---

[92] Op. cit., footnote 83, 348.

[93] St. Thomas, *Summa Theologica*, I, q. 84, a. 7. ['Anyone can experience this of himself, that when he tries to understand something, he forms certain phantasms to serve him by way of examples, in which as it were he examines what he is desirous of understanding. For this reason it is that when we wish to help someone to understand something, we lay examples before him, from which he forms phantasms for the purpose of understanding.']

[94] Lonergan, *Insight*, 464; CWL 3, 489.

[95] F. A. Pareth, 'The Epistemological Status of the Chemical Concept of Element,' *British Journal for the Philosophy of Science* (13), 1962–3, 1–14; 144–150; 8 (original 931).

of physics. Another way of expressing the relation of proportionate matter to form is to note that different types of chemical aggregates give different biological types, and this on the various levels of cell, organ, organism. It would take us too far afield to go into details here but it is worth noting that a general account of genera and species on various levels could be developed on this basis: a series of definite coincidental aggregates of processes, say on the chemical level, would be in one-to-one correspondence with a series of biological laws. This one-to-one correspondence would not be of the type which seems to be implied by Tolman: "'behaviour-acts," though no doubt in complete one-to-one correspondence with the underlying molecular facts of physics and physiology, have as "molar" wholes, certain emergent properties of their own:'[96] the correspondence is rather of the type implied by Bertalanffy: 'nearly homogeneous material—nerve cells, muscle cells, bone cells—builds a typical formation, the outer shape of which is characteristic and specific, while the inner arrangement of the cells is largely accidental.'[97] Thus the various organs are aggregates of cells and on the level of the cell, the aggregate of processes is flexibly different for different types of cell. All this recalls Waddington's remark:

> It is one of the major features of animals—so obvious that its importance is often overlooked—that they are built up of a limited number of rather sharply distinct types of cells ... Animals do not shade off gradually from one type of structure to another, but consist of definite organs sharply distinct from each other ... Development does not produce a continuous spectrum of possibilities, but results in a discontinuous set of different types of cell.[98]

This would correspond to the fact that only particular aggregates of chemical processes are suitable, 'disposed,' 'proportionate matter' for the higher type. This leads directly to the question of the probability of the emergence of such types in a given environment, a question with which we will deal later.[99] Only the principle concerns us here: the principle is that any particular chemical process has a finite probability of occurrence under suitable conditions, that therefore an aggregate of such processes has a finite probability of concurrence under suitable conditions, and the processes occurring thus can be identical with the processes within, say, an amoeba, the only difference being that in the chemical case the total occurrence is non-systematic: in the amoeba the same aggregate of processes is systematic. The transition from non-systematic to systematic is thus associated with the emergence of the higher type.

---

[96] E. C. Tolman, *Purposive Behaviour in Animals and Men*, New York, Century, 1932, 7.

[97] Bertalanffy, *Problems of Life*, 63.

[98] Waddington, *The Nature of Life*, 62.

[99] In the second part of chapter ten.

Our conclusion at the end of this long section on randomness and emergence is fourfold.

First, we have succeeded in specifying further the meaning of randomness and in increasing the evidence for its objectivity. Not only is there the randomness of certain mathematical series, but each particular science involves a new level and type of randomness. Each science therefore provides scope for statistical investigation within it, a statistical investigation which is not entirely reducible through the acquisition of causal knowledge. Thus we have statistical thermo-dynamics at the molecular level, a statistical theory of the distribution of types through time and space in biology, and no amount of classical science can render superfluous the limited generalization or systematization supplied by statistics on these levels. In chapter five we already established this point when we dealt with the interplay and interdependence of the two methods, classical and statistical, in any science.

Secondly, we have associated the problem of objective randomness with the problem of distinct levels of investigation, distinct sciences. In doing so we have strengthened the case for objective randomness and so for the objective reference of statistics—for randomness is its basic presupposition. The case is strengthened because, despite debates in philosophical circles, existence of distinct sciences is in fact the working hypothesis of the scientist *qua* scientist, as it is the implicit assumption of the man in the street. Inversely, however, we have succeeded in giving some theoretical justification for this working hypothesis, thus filling the lacuna noted by Oppenheim and Putnam in their 'Unity of science as a working hypothesis': 'Our working hypothesis rejects merely the claim of absolute irreducibility, unless such a claim is supported by a theory which has a sufficiently high degree of credibility; thus far we are not aware of any such theory.'[100] We would claim that the theory proposed here has that degree of credibility. It is based, not on generalities concerning science nor on a presupposed philosophy, but on the concrete details of investigation at various levels, details considered not so much in their content as in the principles and methods of inquiry they presuppose. Nor does it lose the obvious advantage of the Unity Hypothesis by fragmenting into isolated compartments the distinct sciences. The unity and interplay is preserved because of the type of matter-form relation present between the various levels, lower-level explanation is not excluded but required, and the paradox of excluding higher-level relations and laws in principle while admitting them implicitly in practice is avoided.

Thirdly, we have clarified various aspects of the problem of emergence. The possibility of emergence was seen to be intimately linked with the occurrence of randomness, of the non-systematic, at a given level. Only in so far as there are

---

[100] Oppenheim and Putnam, 'Unity of Science as a Working Hypothesis,' *Minnesota Studies*, Minneapolis, 1962, Vol. II, 15.

present non-systematic processes, such as coincidental aggregates of biochemical reactions, is there a possibility of the emergence of laws, of higher system. The probability of emergence is linked with randomness in so far as randomness is the basis of statistics: it is dependent on the statistical distribution of various types of chemical aggregates, such as those discussed in technical literature on biopoesis, within a given environment. Possibility and probability of emergence are therefore related to randomness in two different ways. Possibility is related to randomness as being non-systematic and therefore systematizable by causal or classical relations at a higher level. Probability is related to randomness as being non-systematic and therefore providing scope for statistical generalization or systematization on the same level. Finally we may add that the discussion of emergence thus far is only the beginning of a clarification and justification of the emergentist view, and of the methodological assumptions involved in any modern principle of evolution. But it is perhaps already sufficient to undermine the view expressed by T. A. Goudge as the conclusion of a survey of emergent evolutionism: 'As for making emergent evolutionism a speculative cosmology or worldview, this is a project fraught with enormous, and perhaps insuperable, difficulties.'[101]

Lastly, the first three conclusions must be qualified by a point made repeatedly already. The relevant meaning of randomness, of emergence, and of their objectivity, will be in the reader's mind only in so far as the reader has been operating in the mode of scientific methodology. Since this mode of reflection lacks acceptability at present, we have preferred to postpone its explicit treatment till the end, thus presenting achievements prior to procedure. Now that the question of emergence has been treated in this way we are in a position to discuss methodological procedure: the paradigm for understanding emergence is in fact central to that procedure. But before the discussion of procedure there are further achievements to be indicated in relation to evolution theory.

---

[101] T. A. Goudge, 'Another look at Emergent Evolutionism,' *Dialogue* (4) 1965–66, 285.

I have an uneasy feeling that some important pieces are still missing from the structure of our theory. I do not know what these pieces are; but I think they have something to do with the fit of what we know about genetics and evolution. We have achieved a synthesis of sorts— but I think that, in the long run, we shall manage to make a much better one.[1]

The present section is concerned with transforming 'an uneasy feeling that some important pieces are still missing' into clues and, further, into a methodological orientation towards the ever-growing data of evolution theory. We might say that our concern here is with 'the meaning of evolution,' yet our approach will differ basically from that of the book by Simpson of this title.[2] The normal aim in books of this latter type is 'to outline in a reasonably clear nontechnical way some major features of the new conception of evolution.'[3] Here, instead of an outline there will be found what may at first appear to be a collection of scattered factors but which in fact are clues pointing towards a particular direction of solution of methodological difficulties. We are not, then, seeking to go beyond the empirical approach of authors like Simpson to some ultimate or 'profound meaning' of evolution. Our concern is the empirical and our quest is for the basic structure of empirical answers. That structure will depend not only on present answers but also on present questions and clues and problems, and thus to a certain extent it will appear to be *a priori*. Yet is it not true that 'the anticipation of Nature is a fraud'?[4] I would distinguish between structure on the one hand and principles, theories, models, and conceptual schemes on the other. *A priori* principles, like those of the Pythagorean School, are vain,[5] theories are replaceable, conceptual schemes can frustrate progress,[6] and 'the price of the employment of models is eternal vigilance.'[7] On the other hand, structure, in so far as it is taken in a fundamental and methodological sense, can be of permanent value. A significant aspect of structure in this sense is that it is both *a priori* and *a posteriori*. As F. E. Crowe remarks

---

[1] Marston Bates, 'Ecology and Evolution,' p. 549 of *Evolution After Darwin* Vol. 1, edited by Sol Tax, University of Chicago Press, 1960. This volume will be referred to subsequently as '*Evolution After Darwin* 1.'

[2] *The Meaning of Evolution*, Oxford University Press, 1949.

[3] T. A. Goudge, *The Ascent of Life*, London, Allen & Unwin, 1961, 14.

[4] R. Harré, *The Anticipation of Nature*, London, Hutchinson, 1965, 108.

[5] Ibid., chapter 4.

[6] Cf. R. Harré, *Matter and Method*, passim; M. Hesse, *Models and Analogies in Science*. Cf. also the reference in footnote 53.

[7] R. B. Braithwaite, *Scientific Explanation*, 93.

Genetically, a method for reaching answers presupposes that we have already reached some of them. Only when we have groped our way down the dark corridor and found the light switch can we look back and see how we should have come. The illustration is rough but will serve, for the corridors of knowing show a pattern.[8]

The pattern, the structure, is revealed only through the consideration of actual investigation, but it can be *a priori* in the sense of anticipating a 'filling-out' through future answers, and in this sense certainly one can class it as having 'open-texture.'[9] In the present case, as we shall see, the 'open texture' is basic and is due to its intimate connection with the classical and statistical procedures of science already described.[10] Moreover, far from being abstract or 'on a high level of abstraction,'[11] as T. A. Goudge claims repeatedly regarding evolution theory, an adequate evolutionary principle or structure in our sense will be, paradoxically, highly concrete.[12] Unlike, say, the general Newtonian differential equation for conic-section orbits, or even its general solution, the passage to the concrete from an evolutionary principle is not just a matter of adding boundary conditions: it includes boundary conditions in an anticipatory way. All this will become clearer as we proceed. Briefly we may say that what we aim at establishing is a fundamental structural classico-statistical principle of evolution theory, 'an approach, a heuristic assumption, that can be worked out in an enormous number of different manners and that can be tested empirically only through such specific determinations and applications.'[13]

We have now expressed, to some extent, what 'meaning' signifies in the phrase, 'meaning of evolution.' But what do we mean here by 'evolution'? Perhaps we might set the stage for explanation by a quotation from J. D. Bernal which we can make our own:

In general the pattern I propose is one of stages of increasing inner complexity, following one another in order of time, each one including in itself structures and processes evolved at the lower levels. The division into stages is not in my opinion an arbitrary one. Although the evolution of life was continuous, for no stage could have been completely static, it cannot have been uniform. Discontinuities which occurred at later stages of organic evolution, such as the emergence of

---

[8] 'The Origin and Scope of Bernard Lonergan's *Insight*,' *Sciences Ecclesiastiques*, IX, 1957–8, 276.

[9] Cf. T. A. Goudge, op. cit., 16.

[10] Particularly in chapter 5.

[11] T. A. Goudge, op. cit., 38, 157, 164.

[12] In what sense this is true will gradually emerge, most especially in the next chapter.

[13] Lonergan, *Insight*, 261; CWL 3, 286.

air-breathing forms, are likely to have been paralleled at the earlier biochemical stages at such jumps as the genesis of sugars, nucleic acids and fats.[14]

It is more in modes of discontinuity and emergence that are similar and repeated in both lower and higher levels of evolution that we are interested, more in what is generic to emergent processes than in what is specific to the evolution of life. But we may profitably push further with the question, what is meant by 'evolution' and how is it distinguished from, say, 'emergence' or 'development'? After a certain amount of reflection T. A. Goudge advances the modest view that 'evolution is a type of change different from mere quantitative increase, individual growth, and individual development.'[15] A more accurate differentiation than this, however, is required and as a preliminary to that differentiation we turn aside to consider the ordinary usage of the words in question.

An examination of origins and usage reveals that the three words 'evolution,' 'emergence' and 'development' have had a checkered career. 'Evolution' has an easy derivation from the Latin with a primitive meaning expressed in the phrase 'unrolling a book.' The dominant usage of the word is for the actual process of unfolding or unrolling but that 'unrolling' passes with ease from the concrete to the figurative meaning so that literally nothing unrolls yet figuratively there is meant the spreading out of a mental vision or the appearance is some orderly succession of a train of events, etc. But between these extremes there is a range of meanings and usages lying closer to the original and which also are more technical. Thus in mathematics alone 'evolution' may refer either to the process of extracting square roots or to the unfolding of a curve, the production of an involute. Again there is the evolution of troop movements or of dance movements. Closer to our own interests there is the usage in reference to the formation of the planetary system. In the biological field 'evolution' refers, as the O.E.D. puts it, to 'the origination of species of animals and plants as conceived by those who attribute it to a process of development from earlier forms.'

Now it would seem that the common meaning running through these different usages is not the original one: the notion of unfolding or unrolling does not lie behind the range of usage in any dominant fashion. The focal meaning in fact would seem to be closer to the meaning of the Latin expression of the scholastics, *eductio,* which they opposed to *creatio.* Something takes shape or form, but that form somehow has its origin within what is taking shape; it is there in potency, and the more that potency was proximate, as the scholastics put it, the more evident was the forthcoming shape. Still, there is a connotation of direction in the usage of the word evolution which is absent in the word *eductio.* For the

---

[14] J. D. Bernal, 'The Problem of States in Biopoesis,' *Aspects of the Origin of Life,* edited by M. Florkin, New York, Pergamon Press, 1960, 30.

[15] Op. cit., 26.

scholastics, chemical forms were educed in the process of organic decay: but no one would be happy to say that they are evolved. 'Evolution' then connotes an absence of symmetry. In what does this absence of symmetry consist? It can be most easily identified, perhaps, as the residue of the primitive meaning of unrolling: to roll and to unroll are not the same thing, and unrolling seems to be a more spontaneous process. Moreover, an unrolling, be it of a scroll or a bud, is a manifestation, a showing of what was latent. One might say that qualities appear that were no more than latently present, that those qualities are somehow an improvement on what was previously there.

To push further here would be to enter into precisions which in fact lie beyond common usage, precisions such as we have already entered into, in chapter nine, e.g., with regard to the direction of evolution as a direction of increasing systematization. But it is worth noting that at the level of common usage the distinction between 'development' and 'evolution' tends to vanish. Thus in the quotation above from the O.E.D. mention is made of the 'processes of development from earlier forms.' Inversely, 'development' is associated in usage with 'evolution, or bringing out from a latent or elementary condition' (O.E.D., under Development). Thus, their ranges of usage overlap and one may speak equally well of the evolution or the development of a civilization or of an economic system. Yet one can detect an important difference, even apart from more technical usages. There is an added precision or stress in the usage of 'development' which seems to go back to its more complex origin. That origin manifests a source of stress or emphasis on unfolding due to the conjunction of the Latin *dis-* with the Roman verb equivalent to the modern Italian *sviluppare,* to unwrap, to disentangle. This stress lends force to such usages as 'the bringing out of latent capabilities' (O.E.D. 3) and 'the gradual advancement through progressive stages, growth from within' (O.E.D. 4). The significance of that stress is neatly brought out by noting the difference between evolving a theme in music and developing a theme in music: there is a connotation of identity in the latter which is absent in the former. To develop a theme one must have the theme already, but normally one is evolving a theme only when one is searching around for the pattern of a melody. Again that difference of stress underlies the changes in usage of the two words. Where earlier the phrase 'evolution of the organism,' meaning the growth of the organism, was not uncommon, it now is. One can take up a modern book on Development in a biology library without fear that it may after all deal with evolution. In current usage, 'development' lays stress on some identity within the process, and in contemporary biology that stress reduces the usage of the word to the field of organic growth. Within that field there still remains the task of making precise the meaning of development. As Paul Weiss remarks at the beginning of his book *Principles of Development,* the question, What is development? seems trivial: 'Does not everybody have some notion of what development implies? Undoubtedly most of us have. But when it comes to formulating these notions they usually turn out to be vague.'[16] Weiss himself seeks to get beyond this vagueness, beyond, too, the type of explanation which 'cannot survive the first rigid test on a concrete phenomenon of development,'[17] by staying as close as possible in his considerations to specific phenomena. Thus, while he sees progressive differentiation as the keynote of development, detailed illustrated discussion of differentiation leaves no room for an accusation of a mere shift of obscurity. Again, the hierarchy of organizations of the organism has to be explained, first by decomposing the complex phenomenon into simple processes of biological order, then further by attempting 'to trace the roots of biological process into the known realms of physical and chemical phenomena,'[18] the ultimate aim being 'to describe and understand any state of the living system as conditioned by the immediately preceding states.'[19]

Weiss' effort may be regarded as a determining of new usage and a precision of meaning of the word 'development.' While that precision is beyond our present scope it is interesting to note how closely Weiss' ultimate aim of being able 'to describe and understand any state of the living system as conditioned by the immediately preceding states,' corresponds to what seems the aim in answering the question, What is evolution? Similarities here, indeed, underlie the tendency among many philosophers of biology[20] to consider evolutionary process as on a par with the process of organic growth, to consider it as the evolution of a macro-organism. But the fact that a similar sequence of conditioned states is involved in both cases is not the whole story. There is a unity of integration and operation in the case of the organism which is recognizably different from that of the total evolutionary process. But to make precise that difference and to give an adequate methodological definition of development would involve a lengthy aside.[21] We turn rather to consider the word 'emerge' which, as we shall see, comes closest to suitably expressing the centre of our attention.

The Latin origin of the word 'emerge' is evident, with the primitive meaning of 'rising out of the water.' That primitive meaning is evident in such a usage as 'accounting for the emergence of these islands,' but the primitive literal meaning is quickly lost by the various processes discussed, e.g., by Stephen Ullmann[22] so that the meaning becomes, e.g., 'the process of coming forth, issuing from concealment, obscurity, or confinement' (O.E.D. 2), 'an unforeseen occurrence; a state of things unexpectedly arising, and demanding immediate attention.'

---

[16] P. Weiss, *Principles of Development*, 1.

[17] Ibid., 75.

[18] Ibid., 108.

[19] Ibid., 120.

[20] Cf. Goudge, op. cit., 32ff.

[21] Cf. B. Lonergan, *Insight*, chap. 8 on the notion of 'thing'; 451–487 (CWL 3, 476–511) on the notion of development.

[22] *Semantics: An Introduction to the Science of Meaning*, Oxford, Blackwell, 1962, 193ff.

(O.E.D. 3). The latter usage of 'emergence' has now been taken over by the word 'emergency.' The former usage can have reference to an unlimited range of things. One may speak of the emergence of teeth, or of the emergence of a race from barbarism. The element of primitive meaning present in all cases is that of 'issuing from concealment, obscurity or confinement' but the concealment is figurative. There is included here a connotation of abruptness which indeed recommends it for use in the development of a more precise specification of evolution. In Ullmann's terms[23] our choice has both morphological and semantic motivation. We note too that while 'emergence' is to some extent transparent—in Ullmann's sense—'evolution' is opaque, contra-motivating in regard to the meaning we would wish it to assume. As G. D. Yarnold remarks;

> It must be admitted that the word, "evolution" is somewhat ill-chosen. Signifying literally an "unfolding" or "unrolling," the word seems to suggest that the final product of the process is somehow contained within its beginning ... Evolution strictly speaking does not explain the present forms of life in terms of the past: rather it traces the emergence of later forms from earlier. While the environmental factor can perhaps be said to operate as a phenomenal cause with a logic of its own, the hereditary factor is characterized essentially as the occurrence of unpredictable novelty. It is this element which renders the word "evolution" somewhat inappropriate; since, accepting the theory at its face value, we may say that apart from the occurrence of novelty there would be no evolution.[24]

'Emergence,' on the other hand, helps to throw the stress properly on the key transition stages. The discussion of the last chapter centred largely on these transition stages, their nature and possibility, and our inclination therefore is to give the single word 'emergence' a dominant position in the quest for scientific precision. Obviously, however, our inclination is no more than that: if it were an occasion of dispute then we would follow the precept of Pareto, 'Never dispute about words.'[25] 'Emergence' and 'evolution' have of course long since come together in the writings of the emergent evolutionists like S. Alexander and C. Lloyd Morgan, but the conjunction of the two words is unhelpful, adding little more than emotive overtone, a thing almost unavoidable except in treatises on logic and mathematics.[26]

This survey of usage, meaning and change of meaning is clearly nothing like a semantic analysis according to standard norms, and the indications of change

---

[23] Ibid., 91–93.

[24] *The Moving Image: Science and Religion, Time and Eternity*, London, Allen & Unwin, 1966, 107.

[25] Quoted by S. Ullmann, *The Principles of Semantics*, Oxford, Blackwell, 1963, 7.

[26] Cf. S. Ullmann, *The Principles of Semantics*, 13.

of meaning are only pointers towards a further change of meaning which we hope to bring about. This type of change of meaning is one that, in semantic studies, does not seem to receive adequate attention. In Ullmann's account of Change of Meaning[27] it is listed among other causes of change of meaning.[28] It is, as Ullmann puts it, the need for a new name because of technological development or, more basically, because of the emergence of new understanding and a new idea or concept. On our view this cause of change dominates many of the other causes. Not that we would underrate these other causes: indeed we would tend to lay greater stress than, say, Ullmann, on the role of the psyche and the organic level in determining usage and meaning and we would consider it a weakness both of analysis and of classical hermeneutics to have understressed that dimension.[29] But our interest here is in the explicitation of new scientific and methodological meanings. If the meaning is novel, there is need not only for a new name but also for a complex of expressions within the language as previously used and commonly and scientifically understood, in order to convey that new name's meaning. The new name and the complex expression will have identity of meaning when properly understood. It is interesting to note the relevance of this line of thought to the paradox of analysis.[30] The root of the difficulty of the paradox of analysis or identity lies in the 'fundamental duality between what may be loosely called "form" and "meaning" or "expression" and "content" as some philologists would have it'[31] for, 'Language symbols are Janus-like: they face two ways. In the Sussurean terminology, they have an external facet, the "*significant*" and a semantic facet, the "*signifie*."'[32] Either one is clear on one's meaning or one is not: if one is clear, then that meaning if novel can be arbitrarily named 'x' and the meaning conveyed by usage and adequate gesture; if the meaning is not clear to oneself, then it can become clear in that very process towards adequate expression. The most evident illustrations of this process occur on the level of logic and mathematics: in both these subjects there is less danger of pseudo-ideas and greater possibility of adequate formulation and of the application of the criteria of Eliminability and Non-creativity which signal identity.[33] In other fields

---

[27] *Semantics: An Introduction to the Science of Meaning*, ch. 8, 193–235.

[28] Ibid., 209; also p. 198, especially d7.

[29] On this question we would refer to the works of P. Ricoeur, esp. *De L'Interpretation*, Paris 1950, and G. Durand, *La Structure Anthropologique de L'Imaginaire*, Paris 1960; *L'Imagination Symbolique*, Paris, 1964. We will touch on this question further in the concluding chapter.

[30] Cf. A. Pap, *Semantics and Necessary Truth*, New Haven, Yale University Press, 1958, ch. 10, esp. 275, 294; also 419–22.

[31] Ullmann, *The Principles of Semantics*, 31; cf. also B. Lonergan, *Verbum: Word and Idea in Aquinas*, 1ff; CWL 2, 12ff.

[32] Ullmann, *The Principles of Semantics*, 31.

[33] Cf. P. Suppes, *Introduction to Logic*, 154.

the identity is more elusive and the danger of the pseudo-idea more proximate, and it is here that analysis finds its justification. Our brief comments above on the primitive meaning and usage of 'evolution,' 'development' and 'emergence' only initiate the analysis in the present case: the relevant analysis is to be identified with the complex of illustrations and refutations which lead up, in the following sections, to the abbreviated formulation of a principle of emergence or emergent probability.

We must consider next in some detail the question, What emerges? What, as Goudge puts it,[34] are the units of evolution? Until recently the explicit answer to this question tended to be 'species,' 'organisms,' 'living things' and such like, or still more recently, various chemical compounds. But there has been a gradual movement away from this approach towards considering populations as the units of evolution. Some such shift is both necessary and advantageous but it gives rise to a variety of problems. As I. M. Lerner remarks,

> It seems obvious that properties of populations, as much as properties of individuals, have evolved under the action of natural selection. At the same time it is not immediately clear as to how natural selection operating on the individual level could lead to the development of integrating factors at the level of population.[35]

The necessity of the wider unit has been viewed from different aspects by different specialists.[36] For example, A. E. Emerson reviews favourably the notion of 'Population as unit,'[37] and points out that such a unit is warranted by competitive and cooperative adaptation, by all the phenomena analogous to flowering plant and pollinating insect, and that 'the unit of selection must be the system composed of both sexes and the young ... It would be extremely difficult to explain the evolution of the uterus and mammary glands in mammals or the nest-building instincts of birds as the result of natural selection of the fittest individual,'[38] that 'death mechanisms are adaptive, not to individual survival, but to group survival,'[39] etc. Darwin of course was by no means unaware of these aspects of evolution, and indeed his writings also give grounds for the suspicion that the population is no happier a unit of evolution than the organism.

---

[34] Op. cit., 26.

[35] I. M. Lerner, *Genetic Homeostasis*, London, Oliver & Boyd, 1954, 4.

[36] One gets a representative sample in *Evolution After Darwin* 1: the different views bear the mark of the specialty of the authors.

[37] 'The Evolution of Adaptation in Population Systems,' 310 ff. *Evolution After Darwin*, 1.

[38] Ibid., 319.

[39] Ibid., 319.

T. A. Goudge touches on some of the difficulties which arise from treating the population as the unit of evolution.[40] There is the difficulty of determining what constitutes a population: is it, for example, the living members, or is it what endures through time, or is it determined by some other type of temporal cross-section? Is its spatial delimitation to be determined by potential infertility, or is it to be identified with the deme, the smallest collection of interbreeding organisms? Later we will see how these questions can be reformulated and answered in an unexpected way. But firstly let us note some further difficulties relating to the notion of the population as unit of evolution.

Whatever way the extent of the population in time and space is determined, or its homogeneity or heterogeneity defined, the discussion of its evolutionary progress tends normally to become a consideration of investigations into genotypes and their variations. One speaks then of a 'gene pool' or of a 'corporate genotype.' According to Dobzhansky the rules governing this population genetic structure are distinct from those governing the genetics of individuals, just as the rules of sociology are distinct from those of physiology.[41] In the previous chapter we already had occasion to criticize the concentration of attention on genotype in the case of the individual organism, and despite differences between corporate and individual genotypes, the like criticism can be levelled at the attitude expressed in the following citation:

> Since inherited traits are controlled by genes, evolution can be redefined as a change in the kinds and frequencies of genes in populations. The problem then becomes to discover how the frequency of a gene already existing in the population may change, or how new types of genes, originating by mutation, become incorporated into the population. In order to study the genetics of a population, it is necessary to consider it, not as a group of individuals, but rather as a pool of genes from which individuals draw their phenotype and to which they in turn contribute their genes to form the pool for the next generation.[42]

Apart altogether from the methodological criticisms levelled at this approach there is the fact that it does not seem to lead to significant results: despite the attention it has received especially from the mathematical point of view, its results are abstract and unrewarding. But we will treat of these later when we come to consider the question of stochastic models.

But our criticism of this approach is not entirely negative. Indeed the criticism has a positive aspect which leads us clearly beyond the notion of 'gene

---

[40] *The Ascent of Life*, 26–34.

[41] *Genetics and the Origin of Species*, New York, Columbia University Press, 1951, 15.

[42] D. J. Merrell, *Evolution and Genetics*, New York, Holt, Rinehart, and Winston, 1962, 234.

pool' or of population as unit. In our earlier criticism of the 'gene' approach we pointed out that gene structure was only a partial determinant of either development or evolution, and this is more evidently true of corporate gene structure. Considerations of gene pool fluctuations and such like are indeed considerations of an extremely abstract process. Being a concrete process, 'no discussion of the emergence of evolutionary novelties can be considered exhaustive which does not include a treatment of the environmental situation. Indeed, most evolutionary changes of structures cannot be fully understood without an analysis of the accompanying environmental changes.'[43] A balance in this matter is gradually being restored through the consideration of the various factors of evolution, a balance which, as Dobzhansky notes, was present in Darwin's appreciation of the variation of the organism in relation to its organic and inorganic conditions of life.[44] This more concrete view is especially due to studies in ecology, a branch of biology which is gradually invading the realm of fossils, until recently a preserve of taxonomists:

> ... But what about the associates of the horse in its developmental stages? What did it eat, and what was its habitat and niche? What were its predators and competitors? What was the climate like at the time? How did these ecological factors contribute to the natural selection which must have had a part to play in shaping the structural evolution?[45]

The paleoecologist attempts to determine from the fossil record how organisms were associated, how they interacted with existing physical conditions, how communities changed with time. Thus the abstract considerations of the statisticians of genetics and the efforts of taxonomists are gradually being replaced by, or better integrated as minor contributions into, the concrete considerations of ecology. Thus if the population as genepool is not a candidate for 'unit of evolution,' the population phenotypically considered is little better. It too is an abstraction: crops depend on sunshine, cattle on rain. Are we then to abandon the search for units of evolution, or more generally of emergence, and to leave to the ecologist the task of determining in particular cases the processes of concrete evolution in a manner somewhat akin to a historical narrative?

We might well at this stage spring to another region, that of history or sociology, for clues to the problem of units. But it seems best to make a start from the precise field under discussion. We turn then for clues to one of the earliest workers on the question of Biopoesis, A. I. Oparin. The following quota-

---

[43] Ernst Mayr, 'The Emergence of Evolutionary Novelties,' *Evolution After Darwin* 1, 367.

[44] Th. Dobzhansky, 'Evolution and Environment,' *Evolution After Darwin* 1, 407.

[45] E. P. Odum, *Fundamentals of Ecology*, Philadelphia, Saunders, 1959, 284.

tion would indeed seem to give an adequate indication of the approach we are looking for:

> . . . when there was rapid and massive growth of the original systems, selection took place, the only ones which were preserved for further evolution being those in which the network of reactions was so coordinated that there arose stationary chains of reactions which were constantly repeated or, even better, closed cycles of reactions in which the reactions always followed the same circle and branching only occurred at definite points on the circle leading to the constantly repeated formation of this or that metabolic product. This constant repetition of connected reactions, coordinated in a single network, also led to the emergence of a property of living things, that of self-reproduction.[46]

And so on. Nor is this an isolated indication: the question of cycles and their relevance recurs throughout the discussion of this chapter.[47] Briefly, Oparin finds that

> any system which could serve as a starting point for the evolution of matter on the way to the origin of life must have been based on the principles of organization is space and time which characterize all living things without exception. As we said above, this condition is fulfilled by a drop of a complex coacervate formed of polypeptides, polynucleotides and other substances of high molecular weight and having the properties of an open system with its characteristic network of reactions which are interdependent in time.[48]

Some form of cyclic process is essential to the constancy of certain sequences of biochemical reactions upon which the mechanism of biosynthesis is based;[49] nucleic acids can only effectively synthesize a protein against such a background of metabolism;[50] nucleic acid itself arises 'on the basis of strictly coordinated, constantly repeated, catalytically induced exchange reactions.'[51] His discussion and argument is abundantly exemplified, so that he can conclude,

> It is clear that no substance which forms a major component of protoplasm can be reproduced by a chance or easily attained relationship between the rates of reactions. It requires the absolutely constant,

---

[46] *The Origin of Life on the Earth*, London, Oliver & Boyd, 1957, 359.
[47] Ibid., chapter 8.
[48] Ibid., 341.
[49] Ibid., 360.
[50] Ibid., 362.
[51] Ibid., 363.

continually repeated chains and cycles of reactions which together comprise the network of the self-reproducing, living, open systems.[52]

Such systems have their origin in the lower-level, dynamically stable, colloidal formations and so on into the level of physics. There is a selection of systems based largely on stability and compatibility, 'destroying those which have an "unsuccessful" combination of reactions and preserving for further evolution only systems with chains and networks which enable them to survive for a long while under conditions of constant interaction with the external medium.'[53]

The approach suggested by Oparin's investigation is one which centres attention not on individuals or populations but on certain types of systems, stable but flexible concretely-operative cyclic systems, or what Lonergan would call 'schemes of recurrence.'[54] A scheme in this sense is a systematic arrangement which includes concrete determinations and the systematic arrangement is one which ensures repetition. Why this is a central clue and how it is of general relevance will be investigated at length. But firstly it is as well to show that what is involved is not something restricted to the biochemical order of evolution or emergence. So, for example, E. P. Odum regularly takes this viewpoint in his discussion of ecology. Here the ecosystem is the focus of attention, but it can itself contain, or be contained in, other cyclic systems. 'The idea of the ecosystem and the realization that mankind is a part of complex "biogeochemical" cycles with increasing power to modify the cycles are concepts basic to modern ecology and are also points of view of extreme importance in human affairs generally.'[55] This obviously is one with his view that the best way to delimit modern ecology is to consider it in terms of the notion of levels of organization.[56] However, not only are there the cyclic processes at chemical and cellular levels emphasized by Oparin, there are also schemes not tied to a particular level. As Odum notes, 'The chemical elements, including all the essential elements of protoplasm, tend to circulate in the biosphere in characteristic paths from environment to organism and back to the environment. These more or less circular paths are known as "inorganic-organic cycles" or *biogeochemical cycles*.'[57] Again, on the interspecific level there is organization around the range of modes of interacting: commensalism, parasitism, predation, etc.,[58] and at a higher or more general level, 'communities have a definite functional unity with characteristic trophic structures and patterns

---

[52] Ibid., 363.

[53] Ibid., 376.

[54] Lonergan, *Insight*, 117ff; CWL 3, 141 ff.

[55] E. P. Odum, *Fundamentals of Ecology*, 26.

[56] Ibid., 5.

[57] Ibid., 30.

[58] Chapter 7 of Odum's book is devoted to this subject. Cf. below, the quotation from Darwin, 222–3, and our discussion of an example from Beckner, 185.

of energy flow.'[59] Nor is this unity statically stable: it passes through several stages to reach a *climax,* a mature community.

But let us try to pinpoint more clearly the basic structure that underlies all these different instances, biochemical, physiological, ecological. We are reminded here of M. Bates' warning: 'The attempt to separate ecology and physiology, to separate what I would like to call "skin-out biology" from "skin-in biology," creates serious difficulties—but so does any attempt to categorize approaches to knowledge.'[60] The advantage of the structure we hope to identify is first of all that it is not a principle of separation of 'skin-in' and 'skin-out' biology—this is probably already evident from our illustrations. Secondly, and more important, the structure is not a category subject to radical revision but a mode of explanation rooted in the classical and statistical patterns of scientific investigation. Our first task will be to throw some light on the classical aspect of this structure, then we will consider the statistical aspect, finally in chapter eleven exposing the 'open-texture' synthesis which unites both by exploiting the notion of scheme-probabilities.

The simplest illustration of a scheme of recurrence is perhaps that provided by the planetary system, or even by the earth orbiting the sun. The first thing to note is that it is a going concern, a concrete realization of the Newtonian—or Einsteinian or such like—law of gravitation, with definite particular conditions and, moreover, easily linked with a representation. In the general case, of course, the scheme involves a combination of classical laws, but even this simpler case can give some notion of how the law is related to the scheme, and also how the scheme facilitates the discovery of the law: one might recall here the history of the discovery in this particular case, from Brahe's observations through Kepler to Newton. It is worth noting in passing how the particular case under discussion illustrates in its origins the emergence of a new scheme, but not of a new law. But we will come later to discuss the problems of emergence and survival of schemes, and their probabilities.

Any other of the schemes of recurrence already mentioned might be examined in detail. It is obvious that these schemes reach an enormous degree of complexity: any number of components may be involved and various alternative routes may be possible. Furthermore, such circular arrangements may be interlocked with others in an interdependent way, or may be complemented by defensive circles as in the case of feedback systems. To develop an illustration of interlocking schemes on the large scale one might analyse the schemes underlying the four major subsystems of the evolutionary system discussed and diagrammed by C. H. Waddington:

---

[59] Ibid., 245; cf. also fig. 11, 47 of Odum's book.
[60] Marston Bates, 'Ecology and Evolution,' in *Evolution After Darwin* 1, 549.

One is the "genetic system" the whole chromosomalgenic mechanism of hereditary transmission; the second is natural selection; a third, which might be called the "exploitive system" comprises the set of processes by which animals choose and often modify one particular habitat out of a range of environmental possibilities open to them; and the fourth is the "epigenetic system"—that is, the sequence of causal processes which bring about the development of the fertilized zygote into the adult capable of reproduction ... We have to think in terms of circular and not merely unidirectional causal sequences. At any particular moment in the evolutionary history of an organism, the state of each of the four main subsystems has been partially determined by the action of each of the other subsystems.[61]

And so on. Cycles of the third system we will illustrate presently. As a process intimately connected with the second system one might consider the well-known Krebs cycle[62] with its concrete sequence of component chemicals, its alternate pathways, its interlocked dependent processes, its determinate reactions in which are embedded a cluster of chemical laws.

Darwin himself was not unaware of such complex schemes, and the following quotation illustrates not only this but also his awareness of the involvement of probability notions, and of the major, or drastic, as well as minor flexibility of schemes.

In several parts of the world insects determine the existence of cattle. Perhaps Paraguay offers the most curious instance of this: for here neither cattle nor horses nor dogs have ever run wild, though they swarm southward and northward in a feral state; and Azara and Renggar have shown that this is caused by the greater number in Paraguay of a certain fly, which lays its eggs in the navels of these animals when first born. The increase of these flies, numerous as they are, must be habitually checked by some means, probably by other parasitic insects. Hence, if certain insectivorous birds were to decrease in Paraguay, the parasitic insects would probably increase, and this would lessen the number of navel-frequenting flies—then cattle and horses would become feral, and this would certainly greatly alter (as indeed I have observed in parts of South America) the vegetation: this again would largely affect the insects; and this, as we have just seen in

---

[61] G. H. Waddington, 'Evolutionary Adaptation' in *Evolution After Darwin* 1, 400; cf. the diagram on 401.

[62] Cf. Loewy and Siekevitz, *Cell Structure and Function*: we will in fact consider this in another context in the next chapter.

Staffordshire, the insectivorous birds, and so onwards in ever-increasing circles of complexity.[63]

The key to the understanding of this complex ecosystem is the scheme of recurrence: one may note, for instance, the intersection of the two schemes, animal birth and egg laying in navel. Again, the repetitive egg-laying scheme is a concrete realization of laws of seeing, flight, laying, etc.

Another biological example which illustrates further this last point, as also the third scheme of Waddington, and also parallels in an illuminating way the earth-orbit example, is that discussed by Morton Beckner;[64] the case of a fish swimming for cover in a hole in a reef when it is attacked by sharks. As Beckner points out, it is a particular case of the generic 'escape-reaction.' Instead of fish and shark one might consider fly and spider, rabbit and dog, etc. According to Beckner, the *explanandum* is the path of the fish. Our approach would be to consider more concretely the activities of fish and sharks. In fact the escape-reaction is recognized only because the process is cyclic, repetitive. What have to be explained in the concrete are the intersecting cycles of animal behaviour: the fish's flexible cycles from safety to open sea in relation to predator cycles. The path of the fish and the identification of stimulus and responses are only steps towards that explanation. They resemble more Brahe's observations or at best Kepler's laws rather than Newton's understanding of the planetary path. To make the parallel explicit one may say that just as the solar system is one concrete realization of Newton's law, so the particular cycle which involves the escape-reaction is one concrete realization of a set of biological laws. These laws are to be discovered by an investigation, not merely of this cycle, but of a whole related range of such flexible cycles of behaviour. The point is echoed in the very different context of sociology by Talcott Parsons: 'It is a fundamental property of action thus defined that it does not consist only of ad hoc "responses" to particular "stimuli" but that the actor develops astern of "expectations" relative to various objects of the situation.'[65] We may profitably illustrate such 'systems of expectations' with an extension of the last example, showing how the expectations are related to schemes of recurrence.

Consider, instead of the behaviour of a fish, the living of a rabbit. If a dog chases the rabbit it will run into a burrow, and come out later when all is clear. Here we have a single instance of a scheme. This scheme has a range—the notion parallels Darwin's idea of range of temperature—endurance of plants, etc.—the range includes the number of available burrows, etc. Again the rabbit has a reproductive scheme, a feeding scheme, etc. There is a circle of such schemes within which the rabbit lives, but the circle is flexible—having escaped the dog

---

[63] *On the Origin of Species*, London, Macmillan, 1910, 63–64.
[64] *The Biological Way of Thought*, 156.
[65] Talcott Parsons, *The Social System*, London, Routledge, 1951, 5.

the rabbit needs neither feed nor sleep. By such a detailed examination one may come to appreciate both Parsons' idea of 'a system of "expectations" relative to the various objects of the situation' and Lonergan's notion of 'a flexible circle of ranges of schemes of recurrence.'[66]

As remarked earlier, we might well have approached this question of units of evolution, and of evolutionary statistics, from more familiar examples or from the social sciences. Thus in ordinary familiar living or in tribal living as the anthropologists find it,[67] it is the concretely-operative structure that survives, and that structure involves a multiplicity of relevant schemes for the recurrence of factors of survival. Breakfast is needed in the morning, but it is also needed every morning. Again, what has survived for centuries at the level of government in England is not a population of deputies or their descendants but a parliamentary system. What gave rise to the depression between the wars was not the absence of components for an economy but the breakdown of a system, a network of schemes of recurrence. The fields of anthropology and sociology would provide more complex and developed illustrations.[68] We restrict ourselves here to the work of Talcott Parsons, already cited. An analysis of the different systems and subsystems he discusses would reveal the presence of various recurrence-schemes. Indeed, his notion of system, or of boundary-maintaining system, parallels closely the suggestions of Oparin with which we began. 'The central point of reference is, as we have consistently attempted to make it, the concept of *system*.'[69] 'The definition of a system as boundary-maintaining is a way of saying that, *relative to its environment*, that is to fluctuations in the factors of environment, it maintains certain constancies of pattern, whether this constancy be static or moving,'[70] and one may note his 'Law of inertia of social process' a conception 'similar to that of homeostasis in physiology.'[71] The tenth chapter of Parsons' book,[72] in which medical practice is defined and dealt with as 'a "mechanism" in the social system for coping with the illness of its members,'[73] illustrates the

---

[66] *Insight*, 460 and 465–6; CWL 3, 485 and 490–91. I have no certainty, however, that my use of the word 'range' here and on 188 coincides with Lonergan's. The inclusion of the Darwinian usage is merely suggestive of degrees of flexibility.

[67] Especially in the tradition of anthropologists whose considerations are more concrete, e.g., M. Mead and R. Benedict.

[68] Cf. footnotes 6, 65, 67. The question of models is relevant here: cf. *The Relevance of Models for Social Anthropology*, New York, Routledge, 1965: D. M. Schneider, 'Some muddles about models; or, how the System really works,' 25–85.

[69] Op. cit., footnote 50, 480.

[70] Ibid., 482.

[71] Ibid.

[72] Ch. X, 'Social Structure and Dynamic Process: the case of modern medical practice,' 428–479.

[73] Ibid., 432.

analysis of a mainly defensive cycle, as does also the chapter dealing with 'deviant behaviour and the mechanism of social control.'[74] And so on.

On page 485 of this book, Parsons contrasts the system, as paradigm, with theory, where the latter is understood in the usual sense as a system of laws. The point parallels to a large extent our own comments on the case discussed by Beckner: the escape reaction. There we noted the relation of the scheme to the laws and to the discovery of laws. Parsons continues:

> To say that we have achieved a paradigm and not a theory is not to say that *no* knowledge of laws is involved ... Without a good deal of such knowledge the paradigm would not be possible. But this knowledge is, relative to the empirical problems to be served, fragmentary and incomplete. The paradigm primarily accomplishes two things. First, it serves to mobilize such knowledge of laws as we have in terms of its relevance to the problems of the explanation of processes *in the social system*. Secondly, it gives us canons for the significant statement of problems for research so that knowledge of laws can be extended.[75]

While Parsons does not explicitly advert to the relevance and cognitional significance of recurrence-schemes, his systems involve essentially such schemes and his remarks on the two things accomplished can be referred directly to these schemes. Consonant with the first accomplishment are the facts already noted about the scheme: that it is a concrete realization of a combination of classical laws, easily associated with a representation, dominated, because it is systematic, by intelligence, being thus, inversely, the pattern of invention. Consonant with the second accomplishment is the fact that the identification of a scheme facilitates the discovery of the laws involved. But as yet we have scarcely touched on this second accomplishment. We will show how 'it gives us canons for the significant statement of problems' by later going beyond schemes to the question of the probability of the emergence and survival of schemes. Parsons' concluding chapter deals with the problem of a theory of change and he remarks that '*a general theory of the processes of change of social systems is not possible in the present state of knowledge.*'[76] Our discussion will lead not to a general theory but to a general methodological structure of change which weaves together the classical and statistical modes of investigation.

The stress so far has been on the schemes as units rather than on their elements or on the precise nature of the recurrence. So, while we spoke of various types of schemes, schemes within schemes, defensive cycles, alternate cycles, etc., we did not touch on the detail of the fluctuations of the elementary parts. Thus

---

[74] Ch. VIII, 249–325.

[75] Ibid., 485.

[76] Ibid., 486: italics his.

in the simple recurrence-scheme of the motion of the earth round the sun, or of the earth round itself, there is no significant variation of the elements involved. On the other hand, biological reproduction as a stage in a complex of recurrence-schemes is unique on each occasion: the pattern is maintained but the components vary. Within the Krebs cycle there are less and more permanent elements, while in the usual economic cycles one can distinguish various degrees of permanence in the elements, from meat to money. The circulation of the blood represents an interesting recurrence-scheme: the channel, so to speak, is relatively permanent compared to the fluid and its content, which changes steadily in transit with a somewhat discontinuous change in the region of the lungs. More complex still are the variations in elements in the various phases of the life cycle of *plasmodium vivax* in mosquito and human, giving rise to a three-day-period malaria.

As is evident from most of the examples, the cyclic or circular nature of the recurrence is not physical but metaphorical. Still, it is a metaphor in harmony with long usage: indeed the majority of usages listed in the O.E.D. are of this type. Neither is the recurrence a well-defined affair, except perhaps in such things as clocks and orbits. Pulse rates vary with stair-climbing and chemical reactions with the temperature. In the case of carbon or nitrogen cycles, the flow of the elements in nature is completely lacking in uniformity so that the best one can do is to consider the period of circulation to be some type of statistical average for each cycle, and, further, average over a series of cycles.

On the level of the organism the complexity of recurrence-scheme leads us to duplicate the classical word 'cycle,' so that we speak of a circle of cycles or a circle of schemes of recurrence. Here however the word 'cycle' is used in a much looser sense. For, the series of recurrence-schemes within which the animal can operate are not so linked that the organism moves in orderly fashion from one to a definite next. Thus, as already mentioned, having escaped the dog—the escape reaction involving various recurrence-schemes (compare Piaget on children, cited below)—the rabbit can move into any of several other schemes of operation. Thus we speak of a flexible circle of schemes. Again, each scheme in which the animal functions has an inner flexibility somewhat like the range of behaviour of Darwin, and this flexibility must be allowed for: the rabbit has no determinate mate, nor has it a determinate burrow to run to, nor has its diet the fixity of a medical prescription. But regularly it eats, mates and runs for cover.

It is only in so far as one attends to the actual process of scientific observation that one comes to appreciate how the behaviour of the animal is patterned in schemes and how the scientific investigation of that behaviour is dominated by the recurrent components of such schemes. Furthermore, in the case of the growing organism there is a further complication: for, the flexible circle of schemes that make up the young organism's behaviour pattern is not identical with—and sometimes, as in frog and butterfly, is extremely different from—the flexible circle of schemes of the adult. Here arises the problem of studying development and of showing the linkage between the members of the sequence of circles of schemes which normally become more complex and flexible with age.[77] One may recall here the work of J. Piaget which deals precisely with this problem at the level of early human development. We may note too the striking parallel between his central notion of *Schema* and schemes of recurrence as we have described them:

> A *schema* is a cognitive structure which has reference to a class of similar action sequences, these sequences of necessity being strong, bounded totalities to which the constituent behavioural elements are tightly interrelated.[78]

> To be sure, it is clear that schemas subsume behaviour sequences of widely differing magnitudes and complexity: compare the brief and simple sucking sequence of the neonate with the complex problem-solving strategies of a bright adult. Schemas come in all sizes and shapes. However, they all possess one general characteristic in common: the constituent behaviour sequence is an organised totality. Thus, an action sequence, if it is to constitute a schema, must have a certain cohesiveness and must maintain its identity as a quasi-stable, repeatable unit.[79]

> One of the most important single characteristics of an assimilatory schema is its tendency toward repeated application. In fact, only behaviour patterns which recur again and again in the course of cognitive functioning are conceptualized in terms of schemas.[80]

It is interesting also to note Piaget's discussion of primary, secondary and tertiary *circular reactions*[81] which involve the repetition of chance adaptations on the part of the infant. According to Flavell,[82] Piaget borrowed this concept from J. M. Baldwin,[83] who had used it in trying to account for the selection and retention of infant habits within a quasi-Darwinian theoretical orientation. The work of Piaget concretely illustrates the nature and importance of schemes of recurrence in the study of human development,[84] and strengthens the case for its heuristic significance in the more general field of evolution.

---

[77] Cf. footnote 21 above, p. 175.

[78] J. H. Flavell, *The Developmental Psychology of Jean Piaget*, 52–3.

[79] Ibid., 54.

[80] Ibid., 55.

[81] Ibid., 92–4 and 101–9.

[82] Ibid., 93.

[83] J. M. Baldwin, *Mental Development in the Child and the Race*, London, Macmillan, 1925.

[84] Aquinas' heuristic consideration of human action and development in the *Summa Theologica*, Ia, IIae, was cast in terms of potency, habit and act. A contemporary Aquinas

would handle the same topic in terms of groups of concrete operations. The relation between the two treatments would be, roughly, parallel to that between laws and concrete schemes which we have touched on more than once.

Perhaps the simplest way to introduce the discussion of the probabilities to be associated with schemes is to consider the most obvious example of an important scheme: the orbiting of the earth round the sun. This scheme clearly fulfils conditions for the emergence of other schemes, but we are interested at the moment only in its own conditions. What are these prior conditions? Obviously the occurrence of the two masses in one of a set of well-defined relations, but also the adequate exclusion of other masses. Only on the adequate exclusion of other masses can the orbit begin to function as a scheme of recurrence. Without that exclusion one might have such conditions as are present in a gas: in that case, while suitable initial conditions for the beginning of a closed orbit might occur, due to collisions and attractions the orbit itself would not be concretely possible. But even under these conditions we can say that the various suitable initial conditions for any particular type of orbit—we recall that we are considering probability as frequency of type of event, not as associated with reasonable betting—have each a finite probability. Consider now the set of suitable conditions for a particular orbit. Under the unfavourable circumstances mentioned the members of the set may be said to be practically unrelated, and so the probability of occurrence of the whole set can best be represented by something like the product of the probabilities of the members. But if the circumstances change—in our simple parallel, something akin to a pressure reduction in the gas—then the set of initial conditions are no longer unrelated. If the earth comes into any one of the set of conditions involving the suitable position and velocity for the particular elliptic orbit, all other members of the set defined by that elliptic orbit will occur. Since therefore, if any of the set occurs, all occur, the change of circumstances would seem to have effected something like a jump in probability from the product to the sum of the particular probabilities. Needless to say, our example is extremely simple and so not representative of concrete problems of calculating probabilities in evolution theory. But at least one can appreciate clearly in this instance what will be less clear and precise in more complex instances, the change in the probability of emergence on the fulfilment of prior conditions or the provision of adequate environment. The notion is very general and heuristic and the possibility of filling out its details remote.

So far we have spoken of probability of emergence: but one may also speak of probability of survival. Thus, in so far as such orbital schemes occur, their continuance in existence is conditioned by a variety of internal and external factors, and so for given circumstances probabilities of survival might be estimated. We recall here that the circularity involved in schemes is not a physical one, but, e.g., a circularity of dependent conditions. So, a stable oscillation under a centre of force can be regarded as a scheme of recurrence.

At the other end of the scale of size a more complex example of the same thing could be provided by a discussion of the occurrence and survival of Krebs cycles. [1] Again there is the need for the realization of the prior suitable biochemical conditions and the presence of the required components before the cycle becomes concretely possible, and once these conditions are realized the probabilities of the component reactions in the cycle tend to complement one another additively. Here, too, there is the problem of the probability of survival of such schemes which leads to a consideration of stability of environments, etc.

All along here we have been concerned with the probability of emergence of schemes and their probability of survival. Certain questions arise spontaneously regarding that exclusiveness and it would be profitable to deal with them in relation to the common Darwinian approach to emergence. What, then, is the relation of the points we have discussed to the origin of species, and to the natural selection of chance variations? The question of chance variations was already considered in chapter six and there the idea that chance variation connoted some objective indeterminacy was excluded. Here we are interested in the positive aspect of the meaning of 'chance variation.' That positive aspect is in fact related to the negative one as von Mises' axiom of frequencies is related to the axiom of randomness. Chance variations do occur with a randomness that excludes predictability in the particular case, but which can be treated in general terms by the use of statistics. The best available instance of this type of treatment is that based on the fruit fly experiments. In such an instance one can clearly see that what is scientifically involved is a statistics of the emergence of variations and combinations of variations. Moreover, what is scientifically relevant to evolution theory are these statistics and not the random divergences from these statistics. These divergences, as we suggested earlier, are much the best candidates for the title 'chance.'

We may consider next the meaning of 'natural selection.' 'The preservation of favourable variations and the rejection of injurious variations, I call Natural Selection' wrote Darwin,[2] and the meaning seemed clear enough. Yet when one questions that meaning a certain circularity appears: the fittest survive, but how could one determine the fittest other than by noting that the fittest are those that survive, and how determine what is injurious other than by noting what is rejected?

Natural selection, which was at first considered as though it was a hypothesis that was in need of experimental or observational confir-

---

[1] Cf. Loewy and Siekevitz, *Cell Structure and Function*, 113–123, with relevant diagrams on 114, 121.

[2] *On the Origin of Species*, 72.

mation, turns out on closer inspection to be a tautology ... it states that the fittest individuals in a population (defined as those which leave most offspring) will leave most offspring.[3]

The slogan is not, however, as barren as Waddington's statement would seem to suggest.[4] The problem is to move towards definite explanation of survival. The suggestions of various authors, to speak not of the fittest, but of the fit, more fit, etc.—or, with Wood-Jones of the survival of the incompletely adapted—point in this direction. While that direction involves developments both in causal and statistical theory, what is most evident is the need for a closely-empirical statistical theory, a statistics of survival. Obviously the elements of these statistics must be equally concretely defined, and this is the other aspect of the problem of fitness: determining the concrete complex of schemes omitting nothing that is relevant:

> The fitness of a genotype or of a Mendelian population has been frequently discussed in terms of the various components of the life cycle, such as fecundity, hatchability, rate of development of immature stages, differential mortality of immature stages, sterility, sexual activity of adults, longevity of adults, etc. It should be noted that all these are components of a single value of fitness and not alternate estimates of it, as some workers seem to imply. Attempts have been made (e.g., Wallace, 1948) to combine these components into *the* fitness in the sense of Wright (i.e. "intrademic selective value"), but one can never be sure that all the components have been included.[5]

One can only move asymptotically towards a sufficient definition of fitness, and that movement requires complementary statistical investigations of the type illustrated already.[6] As a matter of fact, certain variations and combinations of variations 'fit in with the environment in which they occur. That they fit in, and the degree to which they fit in, can be known through statistics of their occurrence and survival. Within such statistics it becomes clear that neither fitness nor survival are a black-and-white affair: a proportion of those that are not the 'fittest'—whatever that may mean—also can survive. Natural selection, then, is not the automatic preservation of some and the automatic elimination of others: it involves the survival of graded types in accordance with probabilities. Darwin's statement, 'The preservation of favourable variations and the rejection of injurious variations, I call Natural Selection,' may then be abbreviated and

---

[3] G. H. Waddington, 'Evolutionary Adaptation' in *Evolution After Darwin* 1, 385.

[4] Cf. Goudge, *The Ascent of Life*, 117 ff.

[5] Francisco Jose Ayda, 'Relative fitness of populations of Drosophila Serria and D. Birchii' *Genetics*, 51, 1965.

[6] In Chapter 5.

rendered innocuous as 'the preservation and rejection of variations,' where preservation and rejection are to be determined statistically.

The last question which we wish to raise regarding the relation of our presentation to Darwin's view is, What of species and the origin of species? We recall here both the question just discussed, Why are certain combinations of variations preserved?, and the four interdependent systems suggested by Waddington:

> At any particular moment in the evolutionary history of an organism, the state of each of the four main subsystems has been partially determined by the action of each of the other subsystems. The intensity of natural selective forces is dependent on the condition of the exploitive system, and the flexibilities and stabilities which have been built into the epigenetic system, and so on.[7]

The survival of variations is a matter of dovetailing, of fitting in operatively with the environment. Now the proximate environment of the variation is the organism, and in so far as one stresses that, one tends to speak of the origin of the species, lethal genes or mutations, individuals or populations as units of evolution, skin-in and skin-out biology, etc. Continuous with this stress is a common attitude of biologists on which Nagel comments:

> When a biologist ascribes a function to the kidney, he tacitly assumes that it is the kidney's contribution to the maintenance of the living animal which is under discussion; and he ignores as irrelevant to his primary interest the kidney's contribution to the maintenance of any other system of which it may also be a constituent.[8]

Our discussion till now in this section has been an effort to remove that stress by focusing attention on recurrence-schemes and systems. Within such schemes the plant or animal may only be a component. This has already been illustrated and it is quite evident on the generic level of sustenance and the birth of offspring. But there is too a large variety of schemes and systems within the organism, ranging from the biochemical cycles within individual cells to the larger schemes such as blood circulation. Moreover these larger inner schemes are usually interlocked with schemes which go beyond the organism. Thus, the vascular circulation depends on the digestive system which depends in turn on the schemes of recurrence by which the organism wins sustenance from the environment. These schemes together with their complexity and flexibility we have already illustrated in the case of rabbit or fish. It is predominantly this flexibility, based on complexity, which enables the animal to survive in a range of

---

[7] *Evolution After Darwin* 1, 400; cf. chapter ten above 183–185.

[8] *The Structure of Science*, 408.

temperatures, pressures, humidity, sunlight etc. Again, this increasing flexibility and complexity is stressed in Darwin's view of the origin of species, but there the stress is more on the capacities of organisms. This increasing complexity and flexibility of organisms in the evolutionary sequence is fairly evident: still, less evident but more significant methodologically is the sequence of schemes of recurrence of growing complexity. The schemes form a sequence, and indeed a conditioned series, in the sense that prior members have to be in operation before the later members become concretely possible. The simplest example of this conditioning is the almost-circular orbit of the earth around the sun at a particular distance, with a particular intrinsic rotation, as a condition for the emergence of life. Again, much of the investigation of this emergence of life is concerned with a determination of proximate conditions on the biochemical and physical level. More obvious examples of conditioning are the dietary schemes of autotrophic and heterotrophic organisms, vegetarian and carnivorous animals.

Before continuing our treatment of the probability of schemes we had best clear up a point already referred to: that our application of probability to the evolutionary process is not in the same category as reasonable betting. When we speak of the probability of schemes or conditioned series of schemes we have in mind the strictly statistical notion of probability. J. Albertson, for instance, fails to appreciate the possibility of this approach when he writes:

> If one should try to speak of the probability of this pencil falling with a certain acceleration when it is dropped now, what is really meant is something describable as subjective confidence in the result, rather than a mathematical probability of that result. From this analysis one comes to the realization that probability can yield no intelligibility for an understanding of the universe as a whole. For by definition the universe ... is a single system, since there can be only one universe; it is not interacting with other systems, because there are no others; and its state is exhaustively described, since one is dealing with the universe as a whole. The only possibility of applying probability theory, then, would be found in an analysis of an indefinitely large number of repetitions in time of the complete universal process—which is another way of saying that probability theory can yield no intelligibility about the universe as a whole.[9]

As we shall see, our application of probability does indeed give some understanding of the universe as a whole, but that application is neither a type of subjective confidence through reasonable betting nor a matter of considering aggregates of complete evolutionary processes. It is a much more prosaic affair,

---

[9] *Modern Schoolman*, XXV, 1958, in a review of Lonergan's *Insight*, 243–4; CWL 3, 268–69.

closely related at all stages to the empirical. As an illustration of what we have in mind we may consider the distribution of what C. S. Coon would consider the five major types, Australoids, Mongoloids, Caucasoids, Capoids and Congoids, on the surface of the earth at any time.[10] Coon's maps for the Pleistocene and early post-Pleistocene concretely illustrate our point. Clearly, the Pleistocene distribution gives a clue to the later distribution: scientifically speaking, on the basis of the actual distribution for the Pleistocene and other relevant factors, a schedule of probabilities could be elaborated for the further stage. This schedule deals with a set of probable next stages. Again, from data on the next stage a similar schedule for a further stage may be elaborated, and so on, through successive schedules of probability for conveniently selected stages.

The principle of evolution or of emergent probability towards which we are moving is, therefore, not some abstract principle of selection for an aggregate of possible worlds. It is geared to the empirical investigation of the successive stages of the actual process of emergence and evolution. In considering the formulation of such a principle it is as well to recall T. A. Goudge's remarks: 'Everybody who writes about evolution needs to be on guard against two dangers. The first is oversimplifying the facts. The second is making unwarranted generalizations,' and again, 'no scientific account of organisms can be satisfactory if it abstracts from their concrete history.'[11] Consider in this light the statement of Simpson: 'Living organisms are all related to each other and have arisen from a unified and simple ancestry by a long sequence of divergence, differentiation and complication from that ancestry.'[12] As Goudge points out, it is the sort of statement that many biologists have in mind when they speak of evolution; it is a statement supported by a large body of evidence and it has no serious competitor in the field. Still, it does involve certain simplifications and assumptions about the relations and origins of living things. Moreover, while it does not definitely abstract from the concrete process of evolution, it speaks directly only of organisms, and that in a somewhat descriptive manner.

Let us consider now an approach which represents what can be called the other extreme in that between Simpson's statement and it there is a wide range of less abstract principles and statistical models.[13] It is the approach of F. M. Fisher, who seeks to produce 'a rather grandiose picture of history.'[14] He defines a *state of nature* as a point in an m-space, m being the number of independent variables excluding calendar time. He introduces time as n discrete moments since

---

[10] Cf. G. S. Coon, *The Origin of Races*, London, Knopf, 1963, especially the maps on 658.

[11] *The Ascent of Life*, 53, 61.

[12] Quoted in Goudge, op. cit., 65.

[13] For a critique of such models, cf. Goudge, op. cit., 42–46.

[14] 'On the Analysis of History and the interdependence of the Social Sciences,' *Phil. Sc.*, 27, 1960, 150.

the creation of the world, and proposes some kind of multiple Markov matrix as giving the required picture of history. 'The typical element of the tensor, then say $Mi_1i_2i \ldots i_{n+1}$, is defined as the probability that Nature will be in the state i_1 at time t_1 given that at time $t-n$ to $t-1$ she was successively in states $i_{n+1}, i_n, \ldots i_3$ and i_2.'[15] Reduction of the number of variables is possible by distinguishing states only on the basis of certain subsets, and so, for example, 'Toynbee's *Study of History* can be regarded as an attempt at a great Markovian reduction of the historical process to a very few variables and very large subdivisions and the consequent description of the process by a multiple Markov tensor of manageable rank.'[16]

Clearly, F. M. Fisher's account of world process is a highly abstract affair, and one moreover which is rooted in Newtonian Mechanics. No doubt this approach may be of value in certain detailed investigations of economics, but as a candidate for a general methodological account of evolution it is a nonstarter. In so far as it does propose a sequence of, say, n schedules of probabilities through time it bears a resemblance to our own considerations of schedules of probability, but whereas the units of relevance for each schedule in our account are to be determined by definition of the actual state of the universe at that stage, Fisher's units are possible states of nature as defined in the m-space.

In general, indeed, as one moves from the type of statement given by Simpson through the range of statistical models, one moves into an undesirable abstraction; one speaks more easily of gene frequencies and fluctuations and less easily of predation, temperature and organic process. Against this background one may better appreciate the need for the introduction of schemes of recurrence into an evolutionary principle. Without that introduction one cannot bring concretely together classical laws and statistical laws, a thing which is clearly desirable, and as we shall see, vaguely anticipated in different generalizations about evolution. Classical laws and definitions of species, chemical elements, etc., have nothing to do with numbers, distributions, time intervals, selectivity, etc. These classical laws, as we saw, in chapter six especially, abstract from particular times and places. Statistical laws depend for their definition of states on classical laws, and give ideal frequencies for the occurrence of events. It does not come within the scope of such statistical laws to explain why there are these kinds of events, why each kind has this particular frequency. To reach a principle which would bring together an anticipation of numbers and distributions and kinds and species and of reasons for the frequencies of kinds, one must include from the beginning a view of classical laws as they are realized in the concrete. The succession of schedules of probability must have for defined units not species, nor individuals, nor populations, nor possible world states, but types of concretely possible schemes. Obviously this introduction of the scheme as unit would

---

[15] Ibid., 149.
[16] Ibid., 156.

complicate considerably the statistics. As G. S. Watson remarks, 'In statistical discussion, it is commonly assumed that the distribution of organisms is uninfluenced by the environment and due entirely to a random, diffusion process with perhaps allowance for births and deaths of organisms.'[17] The common assumption facilitates a high degree of mathematical refinement; the suggested assumption of schemes would render refinement difficult. Still, difficulty is not a criterion of scientific unacceptability. Here we have a situation which is somewhat similar to that encountered when we discussed in chapter eight the problem of adequately defining empirical probability. On a less adequate definition one can erect a complex mathematical structure, but, if empirical relevance is a factor to be considered, attention must be directed to varying empirical conditions and results. In the present case, a complex mathematics has grown up—witness such journals as Biometrics, Biometrika, and the Bulletin of Mathematical Biophysics—on the basis of somewhat abstract evolutionary considerations. This mathematics obviously is of value in considerations of mutations and fluctuations, and would not be wasted in a fuller view. But if one wishes to reach some statistical account of concrete evolutionary process, one cannot afford to abstract from what is essential to that process. On the view exposed here what is essential to the process is a set of sequentially-dependent recurrence-schemes. By taking these schemes, or the series of them—the earlier conditioning the later—in conjunction with their respective probabilities, we reach a view of emergence or evolution as the realization of a conditioned series of recurrence-schemes in accordance with a succession of probability schedules. Finally, as we already noted, the schemes may contain, or be contained within individual organisms. Therefore concomitant with the emergence of schemes there is the emergence of different types of things.

This principle of emergence may seem somewhat trivial. Yet it has a potentiality for explanation which can be illustrated best here by relating it to a range of questions raised by T. A. Goudge in the book already referred to, *The Ascent of Life*. We find it convenient to invert his order of discussion, taking the general problems of purpose, novelty and direction prior to the particular questions regarding the various causal and statistical properties relevant to evolution theory.

We begin with Goudge's last general question, Is there purpose in evolution?[18] The question is clearly worth raising, yet we claim that it lies outside the scope of an inquiry into scientific method. That methodological inquiry, and the principle of emergence which has resulted from it, focuses on the actual procedures of the empirical investigator and the type of explanation he seeks. That type of explanation, as we have already seen, may be to a large extent

---

[17] 'The Distribution of Organisms,' *Biometrics*, 21, 1965, 543.
[18] Op. cit., 191–205.

identified with the Aristotelian notion of form, or again with the formal hypothesis of Aquinas:

> . . . Alio modo inducitur ratio non quae sufficienter probet radicem, sed quae radici iam positae ostendat congruere consequentes effectus. Sicut in astrologia ponitur ratio excentricorum et epicyclorum ex hoc quod, hac positione facta, possunt salvari apparentia sensibilia circa motus caelestes; non tamen ratio haec est sufficienter probans, quia etiam forte alia positione facta salvari possent.[19]

The principle of emergence proposed is a clear illustration of that identification. It says nothing about efficient or final or exemplary causes of the universe: it gives a structural answer to the question of the form, the pattern, of the process of emergence as science knows it and seeks to know it. Like the hypothesis of epicycles which Aquinas talks about as 'saving the appearances,' the present hypothesis makes sense of the empirical data. This we shall see presently in greater detail. It does not make sense of the data by showing the necessity of the process: rather it proposes as a matter of fact that the process has this form. And as Goudge says of Simpson's postulate cited above, so we may say of the present one 'it has no serious competitor in its particular field.'[20]

While we rule out of court the question of purpose, the question of novelty may be treated more positively. At the end of his discussion of novelty Goudge remarks that 'little doubt attaches to the assertion that evolution has indeed generated novelties,'[21] but it remains unclear how novelty might in general be specified. Various authors have considered that there is in fact no novelty, but merely 'reshuffling.'[22] But one should ask further, What type of reshuffling? Is it the shuffling that one associates with card games? Or is it rather the shuffling of the inventor as he gropes about for a new design? One is reminded here of a recent paper entitled 'Design by Natural Selection.'[23] The title serves to stress the fact that questions of purpose or final causes are systematically excluded from our treatment. There are no overtones of this type in our use of words like 'design.' As Simpson remarks in this connection, we are handicapped by language here:

---

[19] *Sum. Theol.* I. q. 32, a.1. ad 2m. ['Reason is employed in another way, not as furnishing a sufficient proof of a principle, but as confirming an already established principle, by showing the congruity of its results, as in astrology the theory of eccentrics and epicycles is considered as established, because thereby the sensible appearances of the heavenly movements can be explained; not, however, as if this proof were sufficient, forasmuch as some other theory might explain them.']

[20] Op. cit., 65.

[21] Op. cit., 167.

[22] Cf. Goudge, op. cit., 166.

[23] By B. Dunham, R. Fridshal, J. H. North in *Form and Strategy in Science*, 306–311.

It is futile for a paleontologist to say, as in effect many have, "By 'orthogenesis' I mean to describe cases in which a trend has gone on for some time without significant deviation; I intend no implication as to why it did so." An odour of finalism still clings.[24]

But thus forewarned we may perhaps legitimately make the comparison that just as the inventor may eventually 'hit on' a viable design, so also may natural selection. The shuffling of the material by the inventor may eventually earn him a patent, a patent which is an acknowledgement of novelty. In evolution the question of novelty is somewhat similar. One must determine in the particular cases whether new processes are such as to call for new laws, or merely for new conditions in addition to old laws, as in the emergence of the solar system. A criterion of novelty would then be the verification of new laws, or for a lesser degree of novelty, the verification of a new pattern of old laws.

The same point may be seen from another angle by considering a prior question of Goudge, Has Evolution overall direction?[25] Here, with Simpson[26] we find quibbles of exactitude unrewarding. There are clearly 'evolutionary traps and blind alleys'[27] and so it is best to stress the word 'overall' in the answering of the question. And in this sense our answer is immediate: the direction of evolution is as a matter of fact a direction of increasing systematization. That increase of system is built into the principle of emergence enunciated above, since prior schemes condition later schemes. Moreover, the notion of conditioned schemes covers both the senses of overall direction which Goudge specifies.[28] There is the amplification of life which we will consider presently in its more detailed expression under the six major features of evolution. There is, secondly, the increase in flexibility, versatility. A point worth noting here is the difference between flexible or homeostatic stability and the type of closed or rigid stability which can exclude incorporation into other systems, such a stability as is manifested in the inert gases or in the non-adaptable extinct species.

Next we come to more detailed considerations of some earlier sections of Goudge's work in relation to the principle of emergent probability. In these sections Goudge makes note of the six major features of evolution[29] certain patterns of evolutionary process,[30] two main groups of items that need explaining,[31] and he gives a sample of ten statements regarding evolution the lawlike

---

[24] *The Major Features of Evolution*, New York, 1953, 268.

[25] Op. cit., 168ff.

[26] Op. cit., 310.

[27] Ibid., 306ff.

[28] Op. cit., 178–9.

[29] Ibid., 38–41.

[30] Ibid., 53–59.

[31] Ibid., 61.

nature of all of which is not evident.[32] Again, he distinguishes historical explanation of two kinds, detailed narrative explanation and broader integrative explanation, and systematic explanation, and there is the problem of their relations. Finally, he raises the question, Is evolution fully explained?[33] We can do no more here than touch on the points of comparison between Goudge's very exhaustive coverage of these different aspects of evolution theory and the present treatment.

We begin by noting a basic aspect of the principle of emergent probability. It is, the manner in which the notion of a conditioned series of schemes involves spatial concentration. This point is brought out well by J. M. Burgers, where his 'accidental circumstances' would correspond roughly either to the basic boundary conditions which, with laws, give initial schemes, or to schemes as conditions of further schemes. It is worth quoting substantially from his account:

> The most pronounced forms of order are directly related to the fundamental laws of nature governing the behaviour of electrons, protons and other elementary particles involved in the structure of atoms and molecules. In other problems of mathematical physics the pattern was dependent, sometimes partially and sometimes completely, upon boundary conditions or initial conditions. In comparison with the fundamental laws we have used the term "accidental circumstances" for these conditions. However, the "accidental circumstances" are the outcome of previous happenings; in other words, they are dependent upon previous forms of order. In this sense we have already spoken of a transmission of order. We have also pointed out that in this scheme of thought the large-scale non-uniformities in the universe must be counted as forms of order; this refers, for instance, to the presence of concentrated sources of energy amidst almost empty space . . .[34]

Now this aspect is centrally relevant to the understanding of the six major features of evolution noted by Goudge. For instance, the gradual increase in structural and functional complexity of living things is heuristically explained by the pyramid-like probability concentration of conditioned schemes. The earlier, less conditioned, schemes have a distribution of probability which favours no particular region overmuch. The multiply-conditioned schemes involved in the operation of animal life have negligible probabilities at earlier stages: indeed, in a matrix of empirical states for a schedule of probabilities of early emergence they would not appear. Only within a later schedule of probabilities can their

---

[32] Ibid., 122.

[33] Ibid., 127.

[34] 'On the Emergence of Patterns of Order,' *Bull. Amer. Math. Soc.*, 69, 1963, 20.

probability of emergence be profitably considered. Thus it is not empirically significant to consider the probability distribution of macro-molecules in the stellar-dust stage of the emergence of the earth, nor to consider human population distributions prior to the Pleistocene. The same point may be made with regard to 'more generalized types of adaptation' and 'the gradual occupation of all possible niches.' For, if there is a type of niche then, on the supposition of large numbers and long intervals of time, one of the sequence of schedules of probability will include the occupation of that type as adequately probable. Evidently too, the occurrence of sequences of schemes and of living things is itself a modification of environment, a creation of new niches, and thus of candidates for these schedules. Steady cumulative changes, speciation and adaptive radiation[35] fit also into the same general structure, involving as they do the statistics of niche occupancy as well as the statistics of mutations. Again, there are the various forms of extinction discussed by Goudge. These fit within the account of emergent probability in that as well as probabilities of emergence there are also schedules of probabilities of survival. Relevant here too are points made earlier regarding fitness and stability: over-neat fitting; over-adaptation can be associated with low probability of survival; over-stability can be associated with high probability of survival but with low probability of integration into other systems.

Regarding the list of ten sample statements about evolution, Goudge asks whether any or all of them can be considered as scientific laws. A classification of these statements is indeed difficult unless one makes use of the fact that explanatory statements about evolution may involve not only laws, but the combination of laws in schemes, the statistics of schemes, and the interrelation of different levels discussed in chapter ten. Thus the statement '(8) Most viable populations exhibit a trend towards increasing specialization,'[36] may be analysed out as a specification of a statistical statement regarding concrete schemes of specialization. The statement '(4) The occurrence of a mutation is always undirected or random,'[37] is a specification of the axiom of randomness which underlies our entire discussion: relative to the biological level, the quite determinate physical events resulting in mutations are non-systematic, random. And so on. The main point to notice is that what is lacking in Goudge's treatment of the different types of statement is a basis of classification, a basis which we consider the principle of emergence, taken in its present context, to provide. A similar lacuna is also evident in the difficulties involved in synthesizing historical and systematic explanations. Goudge rightly remarks that 'no scientific account

---

[35] Goudge, op. cit., 53–59.

[36] Op. cit., 122.

[37] Ibid.

of organisms can be satisfactory if it abstracts them from their concrete history.'[38] This would seem to entail bringing together not only general integrative history and systematic explanation, but the embedding of both in narrative history. The key to the problem lies in adverting to the way in which in the concrete laws are combined with boundary conditions to give schemes. Through the combination of this key factor with statistical considerations one reaches an adequate methodological synthesis of the historical and the systematic. For a further clarification of the nature of that synthesis we can turn to the final question considered by Goudge, is evolution fully explained?

Goudge outlines the state of affairs which would have to exist if evolution were fully explained:

> We may say that at least the following would have to be known: (1) the detailed historical course of evolution; (2) historical explanations of all those single, non-recurrent events which were transitional episodes of major significance in (1); (3) systematic explanations involving generalizations or laws of the various evolutionary patterns or recurrent events in phylogenesis; and (4) the precise manner in which (2) and (3) are to be combined in an overall theory, such that it will account for (1) regarded as a single, complex historical process with large-scale features of its own.[39]

In the first place, we have already indicated 'the precise manner in which (2) and (3) are to be combined in an overall theory': the systematic is to be built into the historical through the device of concretely-operative schemes. Secondly, the combination of statistical and classical investigation-structures required is one based on the fundamental methods of scientific inquiry. It is therefore not a closed model, nor an abstract law or system of laws, but an open-structured anticipation of both classical and statistical laws. Evolution is not fully explained, but the principle of emergent probability is the best possible contemporary anticipation of that explanation.

We can concede, then, Popper's point, that the law of evolution is not a law,[40] and even that it could better be classified as a trend. But where for Popper 'laws and trends are radically different things'[41] for us the trend in question is a clearly defined integration of the two basic types of scientific law. Again, where Popper concludes from the complex intertwining of the laws of concrete

---

[38] Ibid., 61. There are deeper difficulties involved here that we must pass over: cf. B. Lonergan's forthcoming *Method in Theology* on the question of history.

[39] Ibid., 127.

[40] *The Poverty of Historicism*, London, Routledge, 1957, 108.

[41] Ibid., 115–6.

process[42] and the question of testability[43] that 'there are neither laws of succession nor laws of evolution,'[44] we conclude from the same complexity to an evolutionary principle which includes methodically that intertwining and generates testable hypotheses. It generates these hypotheses, not deductively, but by the filling out of the open structure through empirical investigation. In this sense the principle is methodological, anticipatory.

Moreover, the principle represents the anticipation of the best that can be scientifically achieved in accounting for world process. It is nothing like what Mill describes as a method that

> consists in attempting, by a study and analysis of the general facts of history to discover ... the law of progress; which law, once ascertained, must ... enable us to predict future events, just as after a few terms of an infinite series in algebra we are able to detect the principle of regularity in their formation, and to predict the rest of the series to any number of terms we please.[45]

Mill was critical of this method and aspiration, but, as we saw earlier, Laplace, was not. In that earlier discussion of Laplace we based our rejection of his determinist view largely on the analysis of mathematical and physical randomness and the divergence of concrete conditions in physical processes. Since then we have filled out the notion of randomness and illustrated the diverging conditions and the exclusion of system on levels other than that of physics, and the discussion of emergence and emergents in chapter ten enables us to appreciate other limitations on predictability. Laplace was misled by the paradigm of systematic process evident in the planetary system. The concrete process of evolution, however, involves a set of levels of non-systematic or random processes: for such a complex process the granting of knowledge of initial conditions and laws would enable one at best to tackle the impossible task of piecemeal deduction we considered earlier. And the best that one can do towards an integrated view is to establish a schedule of probabilities for the immediate course of events at that stage.

In chapter five, where we discussed causal and statistical explanation we showed how any explanation, causal or statistical, could be presented as a middle term for an Aristotelian syllogism. We consider that the evolutionary process itself could give rise to a similar procedure, but that the required middle term would be a complex mixture of both causal and statistical elements. The formulation of the principle of emergent probability concludes our search for the heuristic form of that middle term.

---

[42] Ibid., 117.

[43] Ibid., 111.

[44] Ibid., 117.

[45] *A System of Logic*, Bk. VI, ch. X, sec. 3.

In chapter nine[1] we pointed out that the relation between, say, biology and biochemistry had to be understood by analogy with insight into phantasm, and there we promised a more adequate account of this claim in the context of a discussion of method. Moreover, the question of method, as we shall see, is intimately related to the meaning of the range of results of the present investigation and so clarification of method will make possible a more precise understanding of these results.

The immediate problem is one of appreciating just how the aggregate of biochemical reactions in the amoeba are related to the set of biological laws verified in the amoeba. With attempted solutions to that problem are associated rules of correspondence, of equivalence, of transformation and such like, and also basic philosophical positions ranging from dualism to reductionism, etc. The problem may be precisely expressed, as we have done, as a question of specifying the relation between two developed sciences, yet it could be raised prior to the development of either science. Indeed, as we shall see more clearly through the solution, the problem does not seem to be essentially changed even when the lower science is ignored. Thus one may have only a vague notion of the chemistry of the amoeba, or no notion at all, and still raise the question of the relation between the chemically-explainable or the merely described or the merely seen amoeba, and the biological understanding of the amoeba. It was indeed in a primitive stage of science that the question was first raised: for the question is closely related to the problem which Aristotle handled in the Metaphysics in his discussion of the meaning of the questions, 'what is a man?,' 'what is a house?'[2]

The close relation of the two problems becomes clear by following the lines of the Aristotelian solution. The clue to the meaning of such a question as 'what is a house?' lies in the division of questions given in the second book of the Posterior Analytics. That division was fourfold: either one asked (a) whether there is an X or (b) what is an X or (c) whether X is Y or (d) why X is Y. Now these questions fall into pairs: (a) and (c) are the types of question which, as we have already seen, are to be associated with verification, with the arrival at an answer Yes or No.[3] Less evidently, (b) and (d) form a pair. Thus in terms of Aristotle's stock example, 'What is an eclipse of the moon?' and 'Why is the moon thus darkened?' are not two different questions but basically one and the same. In more modern terms, 'Why does light refract?' and 'What is refraction?' ask the same question. The answer to either of the latter pair of questions will be an adequate theory of refraction.

---

[1] See 165ff.

[2] *Metaphysics*, z, 17, 1014a, 9ff.

[3] Chapter 7, 110 and passim.

We have already to some extent anticipated our discussion here by associating the modern usage of 'theory' with the Aristotelian usage of 'form,' but we wish now to push the point further.[4]

> Granted that we know what is meant by "What is *X*?" when the question can be recast into an equivalent "Why *V* is *X*?" yet one may ask, quite legitimately, whether there always is a *V*. It is simple enough to substitute "Why does light refract?" for "What is refraction?" But tell me, please, what I am to substitute for "What is a man?" or "What is a house?"[5]

This is the problem which Aristotle tackles in the Metaphysics.[6] There the meaning of the questions 'What is a man?,' 'What is a house?,' becomes 'Why is this sort of body a man?,' 'Why are stones or bricks arranged in a certain way, a house?' What is revealed by the switch in the structure of the question in these cases is the duality of data and anticipated understanding. Indeed, a more refined reflection would show that the three elements of the question, 'Why,' 'this sort of body,' 'a man,' can be related as the terms of a scientific syllogism.

> The Aristotelian formulation of understanding is the scientific syllogism (*syllogismus faciens scire*) in which the middle term is the real cause of the presence of the predicate in the subject. But the genesis of the terms involved in scientific syllogisms follows the same model: sense provides the subject, insight into sensible data the middle, and conceptualization the predicate, which is the term whose genesis was sought.[7]

But it is the more elementary relation of data to understanding that concerns us here and its role in determining the meaning of the terms 'matter' and 'form.'

For Aristotle, the general answer to questions of type (d) was, 'because of the form.' But in admitting this Aristotelian answer 'because of the form' to the why-question it is essential to avoid misconception. One may do this conveniently by emphasizing that the word 'form' is introduced merely to name the answer to the question 'why?' with respect to the data. For, form is uniquely defined by the analogy, form is to matter as understanding is to presentation. Evidently, matter is defined by the same analogy. Matter and form are two interdefining terms, their definition arising from an understanding of the relation of understanding to presentation, or the relation of what-answers to data. Moreover, the evident looseness of these sources of definition points to the flexibility of the analogy, a flexibility which allows it to cover the various expressions of the problem with

---

[4] Cf. chapter five, in particular 75.

[5] B. Lonergan, *Verbum: Word and Idea in Aquinas*, 12–13; CWL 2, 27.

[6] *Metaphysics*, z, 17, 1014.

[7] Op. cit., footnote 5, 15; CWL 2, 28.

which we began this chapter. Understanding can be either of the sensibly perceived or of the imaginatively represented. As the term 'form' covers what is reached by such understanding, so its correlative, 'matter,' covers the counterpart of that understanding. But let us see that analogy operating in a modern context and in the solution of our original problem.

Consider the problem of understanding the amoeba. That problem can be posed in two ways, depending on the state of development of the sciences or the particular scientist's understanding. At an initial stage one may observe the amoeba, an observation which can be refined through the development of various techniques and microscopes. In this case there can be understanding of what is sensibly presented, this duality of understanding-sensing being an instance of the pair form-matter. But one can also approach the problem of understanding the amoeba against the background of physics and chemistry that we discussed in chapter nine. Most evidently, this advance into physics and chemistry represents a movement from description to scientific explanation. But what is important in the present context is the cognitional difference which corresponds to a transition from sensible presentation to symbolic representation. Clear illustrations of such transitions are most easily found on the level of mathematics. Thus a cart wheel or the sketch of a round plane curve can lead to the question, Why is this round?, and to the definition of a circle. One is dealing here with sensible presentations. But the image can be less adequate than this, as when one draws two straight lines and supposes them to be parallel. The important instance of presentation for us, however, is the symbolic presentation in which the symbols stand for whatever one assigns them to stand for. The simplest illustration of this type of image is the drawn or imagined straight line which can neither be drawn nor imagined without breadth but which stands for or is thought of as without breadth. But the illustration most relevant to our purpose is that which occurs in discussions of the amoeba against the background of physics and chemistry. Here, instead of operating with the sensibly-presented amoeba, the scientist operates in the context of symbolic images of the chemical and physical processes. Examples of the procedure were in fact given in chapter nine. The process, indeed, is familiar to the scientist in this field; he makes use of it in textbooks, in teaching, in research; but he does not make explicit, as we try to do here, what precisely is going on. Instead, then, of the microscopically-observed amoeba he has the image of the coincidental aggregate of chemical or cytological reactions, and as we saw at some length in chapter nine, the biological relations provide him with a systematic understanding of these coincidental aggregates.

So we come to see that the basic analogy, form is to matter as understanding is to presentation, leads to an answer to our initial question regarding the relation of the biological to the biochemical. In cognitional terms, the symbolically-presented biochemical level is to biological understanding as matter to form.

All along here we have shown a preference for cognitional terms, but that preference should not be taken to mean that the realm of theory and explanatory hypotheses is purely subjective. Our position is a realist one which grants to the verified explanation an objective status. The realism involved might conveniently be called structural realism. In Aristotle's handling of the problem involved in the question 'Why is this a man?' one can discern the beginnings of that structural realism. There is evidently structure: understanding is conceived as related to presentation. But there is also a realist assumption: understanding, when correct, is not just subjective. Implicitly then in Aristotle there is an assumption of an isomorphism between the structure of knowing and the structure of the real. As verified understanding is to phantasm, so the real form of the thing is to its matter.

> Aristotle's basic thesis was the objective reality of what is known by understanding: it was a commonsense position inasmuch as common sense assumed that to be so; but it was not a common sense position inasmuch as common sense would be able to enunciate or even to know with any degree of accuracy just what it means and implies.[8]

Within the context of the writings of Aquinas, and more clearly within the context of contemporary science, that structure of knowing can be better appreciated as including the component corresponding to verification. In Aquinas' terms, not only is there form and matter: there is also existence. In the terms of modern science, not only is there a theory about the phenomena required: it is required also that it be verified. Knowing is thus triply-structured, and it is assumed that the facts reached by knowing are isomorphically triply-structured. Here then we have a type of correspondence theory of truth. But it is not the correspondence of the structure of propositions to the structure of facts: the correspondence is of structured knowing to the known.

But let us reflect further on the precise nature of these conclusions and how they have been reached. Since part of the conclusions is about knowing, part of the question leading to those conclusions is, What is knowing? It is illuminating to switch this question into the other Aristotelian form, Why is X knowing? X in this case designates the data, which is the experience of knowing, and Why? asks for the form, the structure of knowing. The question is about the process of knowing and the answer is forthcoming through attention to oneself in that process. Thus, one appreciates the difference between seeing blue and understanding blue, seeing a dog and understanding a dog, by reflecting on one's experience of that difference. Again, one may appreciate the difference between understanding and understanding correctly by a like reflection, and the development of modern science with its clear distinction between theory and

---

[8] Ibid., 20; CWL 2, 33.

verified theory makes the reaching of such an appreciation easier for us than it could have been for Aristotle.

The reflection in question here is empirical and scientific in a generalized sense, its data being the activity of scientific knowing in oneself, and the understanding one reaches of that scientific understanding can be constantly checked against one's actual performance of scientific knowing. So, for example, it makes sense to ask the question, Is it a fact that scientific knowing is a matter of verifying theories in particular instances? The answer is reached, not by some peculiar looking in at oneself, but by attending to oneself in the process of scientific knowing. Self-attention undoubtedly seems an odd suggestion for the philosophy of science. As J. L. Synge remarks, in commenting on Hadamard's reflections on the working of mathematicians' minds,

> such things may strike us strange and rather fascinating, a strand of queerness enlivening the dull desert of scientific thought, arid stretches of logic. We may dismiss them lightly and pass on to the serious consideration of what thought and understanding are in terms of the words that philosophers have been accustomed to use. But we may be quite wrong in this. We may miss the turning leading to an under-standing of understanding.[9]

Indeed, there seems no other way to turn if one wishes to appreciate the method of science: for, science is human knowing, the method of science is a process of human knowing, and so the data for the understanding of scientific method lies essentially within the subject's own processes of knowing. One may counter this by saying that one must rather reflect on reality and the facts of science. But it is only a myth which would grant that such reflection somehow goes outside the processes of knowing, that the real order can be reached other than through human knowing, that in some way knowledge can be compared with reality. Instead of such myth we have the assumption of the isomorphism of the structure of the real with the structure of knowing and, as we pointed out earlier,[10] that assumption is unavoidable, inescapable. Any effort to avoid that assumption implicitly assumes it.

In a recent article on 'Methods of Inquiry,' T. E. Burke concludes that 'every enquirer must assume the reliability of some method or methods of enquiry and it is, in principle, impossible for him to produce any arguments in support of this assumption.'[11] He goes on to claim that if two thinkers, such as Descartes and Hume, differ in their basic method of inquiry, then if their particular conclusions

---

[9] J. L. Synge, *Science, Sense and Nonsense*, London, Cape, 1951, 112.
[10] Chapter seven, 119.
[11] *Mind* (LXXIII) 1964, 542.

differ they have no alternative but to simply agree to differ. Relativism is inevitable.

Our present discussion makes explicit our previous indication of the manner in which such relativism can be overcome. The basic assumption of isomorphism exhibits a fundamental implicit presupposition of all inquiry and argument. Hume and Descartes undoubtedly had different theories about what human knowing was and what was real: but what was common to both of them, and in performative contradiction to their explicit conclusions, was the intelligence and reasonableness with which they proposed and defended their positions as true. Both implicitly assumed that intelligent argument and reasonable judgment would lead to a determination of what is the case, and that assumption is equivalent to our assumption of isomorphism. This is so because both men mean what they say, and they mean that what they say is the case. Their meaning is, so to speak, Aristotelian meaning, arising as it does from their understanding of the given. Their meaning may be wrong, but it is nonetheless doubly-structured. That doubly-structured meaning could be crowned by an affirmative answer to the question, Is it the case? Hence, to Descartes and Hume and others who reject the structural correspondence view one may apply Lonergan's comment on the work of L. Dewart:

> ... what is meant may or may not correspond to what in fact is so. If it corresponds, the meaning is true. If it does not correspond, the meaning is false. Such is the correspondence view of truth, and Dewart has managed to reject it without apparently adverting to it. So eager has he been to impugn what he considered the Thomist theory of knowledge that he overlooked the fact that he needed a correspondence view of truth to mean what he said.[12]

Briefly, if the assumption that what is real is what is intelligently affirmed to be the case is to be challenged, it is to be challenged by an intelligent alternative or an intelligent criticism and the reasonable conclusion, 'therefore it is not so' : inevitably, then, the challenger assumes the position he challenges.

Because of the irrefutable and inevitable basic assumption of isomorphism, a reliance on an appreciation of the structure of knowing for an appreciation of the structure of the known is equally inevitable. With this in mind we may now profitably return to the conclusions reached in the entire work and express them with a more direct cognitional stress.

Thus, randomness may be precisely defined in relation to the exclusion of a certain type of understanding. If one has systematic understanding of a process

---

[12] 'The Dehellenization of Dogma,' *Theological Studies* (28) 1967, 338. Republished in *A Second Collection*, edited by William F.J. Ryan, S.J. and Bernard J. Tyrrell, S.J., London, Darton, Longman and Todd, 1974, 14–15; CWL 13, 15.

or a situation then one has something like the formula for an orderly series or for a planetary orbit. The process or series is grasped as a whole in a unified fashion. Such systematic understanding is easily distinguished from the piecemeal fragmented understanding, for example, of the movements and collisions of ten billiard balls discussed in chapter three. The partial understandings of each part of the process in such a case have only unity through the spatial unity of the process: they are an unsystematic or random aggregate. In this fashion one distinguishes, in direct cognitional terms, systematic understanding from non-systematic understanding. Again, one makes more explicit the cognitional counterpart of the problem of real randomness by asking whether the development of correct human understanding could lead to the elimination of non-systematic understanding. The question of real randomness can indeed have no other genuine meaning. For, randomness is defined in relation to human knowing, the real is defined in relation to correct human knowing, in relation to an affirmative answer to the question, 'Is it so?,' and so the exclusion of real randomness can only be discussed in relation to the development of correct human knowing. The question, therefore, is not about some nonhuman intelligence. It asks rather, Can the human scientist hope to reach a systematic understanding of all that he can know? If he cannot, then the affirmation of real randomness is warranted. If there appears to be a jump here between the cognitional and the real, it is merely the jump of the basic assumption of isomorphism. On that assumption the question, Is there real randomness? is identically a question regarding human knowing, the question, Is human knowing inescapably non-systematic? Again, that question is about human knowing as it actually is, and the conclusion is *a posteriori*. But because the conclusion is *a posteriori* it does not follow that the positive answer to the question is only a temporary conclusion. For, the development of human knowing makes possible an appreciation of its structure, and in so far as elements of that structure are invariant they make possible conclusions which not only regard present knowing but characterize invariantly human knowing. The conclusion to real randomness is in this sense a conclusion about human knowing, present and future.

In similar fashion conclusions were reached regarding statistics and emergence. Our discussion of statistics in the earlier chapters was in fact clearly cognitional, with a stress on the type of understanding involved in that field. If the initial treatment of emergence in chapter nine was less clearly so, the discussion of the basic analogy in the beginning of this chapter has remedied the defect. Again, while the consideration of schemes of recurrence was in the context of the problem of real units of evolution, the cognitional significance of such schemes was also stressed. Cognitionally speaking, the scheme can be thought out and to some extent imaginatively represented with facility; it brings scientific laws out of their abstractness into the region of secondary determination, and so it can fit into a statistics of concretely emergent processes.

It is, however, the ontological significance of the recurrence-schemes that is emphasized in the works, say, of Oparin on biopoesis or Odum on ecology. Finally, the effort to formulate a principle of evolution was explicitly made in cognitional terms. What was sought was a structured anticipation of the ever-increasing scientific understanding of the data of evolution. But, to recall the assumption of isomorphism, that structure is also the structure of the real process. In less exact but helpful terms one can say that the structure of the anticipation, which becomes more refined through the development of science, is imposed by the structure of the world known by science. In this case, as in the others, our conclusion can be cast entirely in cognitional terms and the basic assumption of isomorphism ensures that the structure is valid of the object known and to-be-known.

So the conclusions reached here about randomness, statistics and emergence are conclusions regarding both human knowing and the real world, but these conclusions were reached uniquely through self-attention in the process of doing the relevant sciences. We have tried to make clear the significance of the method and its relation to the Aristotelian position, and in that light the entire work can be appreciated as a step towards solving the problem of philosophic method. We may further that appreciation by relating the method to some contemporary views on philosophy, thus returning to the concluding questions of the first chapter.

The questions which led to our conclusions were all specifications of the general question, What is knowing? This, to use Wisdom's phrase,[13] is a basic philosophic stimulus, and in our case it was specified to such questions as, What is systematic knowing?, What is statistical knowing?, What is knowledge of Evolution? In the paper referred to, Wisdom remarks on the ambiguities of the expression 'What is ...?' and points out that

> the philosopher is not using "What is ...?" in the scientists' sense. When the philosopher asks 'What is a chair?," "What is water?," he is not asking for the chemical formulae for these. This may seem too obvious to be worth mentioning. But there are people who speak as if it is the business of the philosopher to carry the work of the scientist a stage further.[14]

But what could that further stage be?

> "What is water?," asked with profound look and in the philosophic manner, has a confusing verbal similarity with "What is water?" asked briskly and in a scientific manner. But the two requests differ in kind,

---

[13] 'Ostentation,' in *Philosophy and Psychoanalysis*, Oxford, Blackwell, 1953, 1–15; reprinted from *Psyche* (XIII), 1933.

[14] Ibid., 13.

not merely in degree. When has a philosopher ever pushed a scientific inquiry a stage further?[15]

Wisdom goes on to eliminate various possible goals of the philosopher's quest, various types of answer to the question 'What is ...?' Ultimately he concludes that the philosopher seeks a clearer apprehension of the arrangement of the elements of the fact, the structure of the fact located by such a sentence as 'aRb.'

Wisdom's effort to specify the philosopher's quest as a clear apprehension of the structure of the fact resembles in statement our own, but we have attempted to specify accurately how precisely that clear apprehension is attained, what precisely the structure of the fact is. Let us return to the question, 'What is water?' As Wisdom remarks, that question can be put with 'a profound look.' Put in that fashion we would consider it to be a quest for an artistic or symbolic answer such as 'Water is the eye of a landscape.'[16] In so far as philosophy is not a well-defined clearheaded pursuit, something like that answer can, of course, come within a philosophic tradition. The second mode of asking the question, 'What is water?,' is strictly scientific. But there is a third manner of asking the question which is neither artistic nor chemical but which regards the other two types of question and answer. That mode is a specification of the basic question, 'What is knowing?' As we have seen, that question can be more explicitly put as, What is the structure or process of knowing? One gives it the required specification by adding the particular object of interest. In the present case the question becomes, What is the process of knowing water? The question then may regard the method of chemistry or of the chemist, and the answer will be reached through generalized empirical method as we have described it. But the methodological question can also be put in relation to the artistic question and then one must extend the considerations beyond knowing to affectivity to reach relevant canons of hermeneutics and methods of symbolic investigation.[17]

The procedure in both cases can meet the demand contemporarily made for clearheaded philosophy. It gives no new information within the science or thought pattern investigated, but makes possible, for example, a methodological ordering of content, thus leading to clarification and the exclusion of nonsense. But it goes further than clarification. As A. H. Price remarks,

What the consumer mainly needs, I think, is a *Weltanschauung,* a unified outlook on the world. This is what he is asking for when he asks the

---

[15] Ibid., 13.

[16] The extent of the artistic, and therefore the dimensions of its scientific and heuristic investigation, should not be minimized. The scientific pattern is the rare, the dramatic pattern is the common one of human experience. One may recall here the treatment of water in the works, say, of Eliade, Durand, Bachelard.

[17] Cf. G. Barden, *Some Strategies in the Investigation of Symbolic Meaning* (Oxford B.Litt. Thesis, 1967; Dept. of Anthropology).

philosopher for wisdom or guidance, or a clue to "the meaning of the universe"; and this is what the analytic philosophers are failing to give him.[18]

Price's own view is that there is not only analytic clarity but also synoptic clarity.[19] Now synopsis normally connotes abstraction, whereas one would hope that an adequate *Weltanschauung* would not be abstract. There is a way out of this dilemma in so far as one requires that the synopsis be heuristic, methodological. For then the synopsis can concretely anticipate all there is to be known. The *Weltanschauung* thus given is not a set of abstract propositions or a speculative metaphysics, but a structured anticipation. Moreover, that anticipation may not be the method-ological anticipation of the results of just one science, but an integrated anticipation of the results of a hierarchy of sciences, such indeed as our inclusive principle of emergent probability provides. Finally, as we have emphasized throughout, the anticipation is neither present nor appreciated in an *a priori* fashion: it develops with the development of the sciences, and the appreciation of it grows with the development of the science of methodology. The anticipation has, indeed, as we have seen, a basic structure which is operative even in the most elementary instances of knowing, and such a basic structure would satisfy Strawson's search for structures of permanent validity.[20] The universal validity of that elementary structure is brought out from a somewhat different angle by F. E. Crowe in an article entitled 'Neither Jew nor Greek, but one Human nature and operation in all.'[21] There he shows that, however different the interests and pursuits of the early Jews and Greeks, those interests manifest in both groups a common structure of inquiry, the structure we characterized in terms of what-questions and is-questions. Again, cultural differences between East and West are a matter of common acknowledgement, but 'when an Easterner inquires and understands, reflects and judges, he performs the same operations as a Westerner.'[22]

Once one moves beyond this elementary aspect, however, Strawson's view that 'the central subject matter of descriptive metaphysics does not change'[23] requires qualification. For, knowing and the structures of knowing develop with use. To neglect that development would be to open the way to a discouraging fixity of metaphysics if not to a type of deductivism. Certainly were metaphysics

---

[18] H. H. Price, 'Clarity Is Not Enough,' in *Clarity Is Not Enough: Essays in Criticism of Linguistic Philosophy*, ed. H. D. Lewis, London, Allen & Unwin, 1963; reprinted from *Proc. Aristot. Soc.* Supp. XIX, 35.

[19] Ibid., 39.

[20] Strawson, *Individuals: An Essay in Descriptive Metaphysics*, 9.

[21] *Philippine Studies* (13), 1965, 546–571.

[22] B. Lonergan, *Insight*, 736; CWL 3, 758.

[23] Strawson, op. cit., 10.

to be restricted to basic and general invariants, then its relevance to the complex issues of developing science and culture would be limited and decreasing. As we have conceived it, metaphysics can grow with the growth of science and culture and it can express the structures of contemporary thought in a refined heuristic fashion. Ogden and Richards remarked in *The Meaning of Meaning* that 'it is not always new words that are needed, but a means of controlling them as symbols.'[24] The metaphysics envisaged here inevitably adds new names in so far as the understanding of structure of knowing and of the known are better understood, but it adds no new content to the science investigated. What it adds is the possibility of the control of meaning envisaged by Ogden and Richards. Control and clarity of meaning have undoubtedly been a centre of attention of analytic philosophy for many years now. The present effort goes beyond the analytic effort in giving closer attention to all that is contemporarily known through common sense and science as it emerges in the mind of the knower. The close attention involved was precisely specified as self-attention, and the result of that self-attention is an appreciation of structured anticipations of what is to be known, ranging from the simple triple structure of every scientific effort to an intricate contemporary anticipation of an understanding of world process.

---

[24] *The Meaning of Meaning: A Study of the Influence of Language Upon Thought and of the Science of Symbolism*, London, Routledge, 1930, 20.

Index

Index

Index

Wisdom, 212, 213
Woodger, J. H., 63, 73, 160, 164
world process, ix, xiii, xxv, xxix–xxxi, xlii–
   xliii, xlvi–xlvii, lxiv, 29–30, 79–81, 94,
   105, 175, 197, 204, 215

Yarnold, G. D., 176
Yule, G., 11

Philip McShane (February 18, 1932–July 1, 2020) was an Irish Canadian mathematician, philosopher, economist, and theologian. He earned an M.Sc. in relativity theory and quantum mechanics with First Honors from University College, Dublin (1952–56), where he lectured in mathematics before doing his D.Phil. at Oxford (1965–68).

Once describing himself as "a dabbler, a mathematician gone astray, rambling in the worlds of economics and literature, music and physics," McShane published works ranging from the foundations of mathematics, probability theory, and evolutionary process to essays on the philosophy of education. He also wrote introductory texts focusing on critical thinking, linguistics, and economics. In the area of methodology, some of McShane's book are *Randomness, Statistics and Emergence* (1970), *The Shaping of the Foundations* (1976), *Lack in the Beingstalk* (2006), and *Futurology Express* (2013), and in the area of theology *Method in Theology: Revisions and Implementations* (2007), *The Road to Religious Reality* (2012) and *The Allure of the Compelling Genius of History* (2015). Among his introductory works are *Wealth of Self and Wealth of Nations* (1975), *A Brief History of Tongue* (1998), and *Music That Is Soundless* (2005, 2nd edition).

For over fifty years McShane was profoundly influenced by Bernard Lonergan's (1904–1984) major works in economics and his break-through discovery of the dynamics of global collaboration. Many consider McShane the leading interpreter of Lonergan's *Insight: A Study of Human Understanding*, a compendious work that lays out both a genetic method for studying organic development and canons for a methodological hermeneutics.

In the last years of his life, McShane wrote with increasing clarity about the negative Anthropocene age in which we live and a future positive Anthropocene age of methodological luminosity and glocal collab-oration. In *Economics for Everyone* (2017, 3rd edition), he indicated crucial steps for seeding the positive age when the "cultural overhead" of leisure will be understood, taught, and practiced, thus freeing many and increasingly all to pursue activities leading to genuine human development.